Interdisciplinary Systems Research
Analysis — Modeling — Simulation

The system science has been developed from several scientific fields: control and communication theory, model theory and computer science. Nowadays it fulfills the requirements which Norbert Wiener formulated originally for cybernetics; and were not feasible at his time, because of insufficient development of computer science in the past.
Research and practical application of system science involve works of specialists of system science as well as of those from various fields of application. Up to now, the efficiency of this co-operation has been proved in many theoretical and practical works.
The series 'Interdisciplinary Systems Research' is intended to be a source of information for university students and scientists involved in theoretical and applied systems research. The reader shall be informed about the most advanced state of the art in research, application, lecturing and metatheoretical criticism in this area. It is also intended to enlarge this area by including diverse mathematical modeling procedures developed in many decades for the description and optimization of systems.
In contrast to the former tradition, which restricted the theoretical control and computer science to mathematicians, physicists and engineers, the present series emphasizes the interdisciplinarity which system science has reached until now, and which tends to expand. City and regional planners, psychologists, physiologists, economists, ecologists, food scientists, sociologists, political scientists, lawyers, pedagogues, philologists, managers, diplomats, military scientists and other specialists are increasingly confronted or even charged with problems of system science.
The ISR series will contain research reports — including PhD-theses — lecture notes, readers for lectures and proceedings of scientific symposia. The use of less expensive printing methods is provided to assure that the authors' results may be offered for discussion in the shortest time to a broad, interested community. In order to assure the reproducibility of the published results the coding lists of the used programs should be included in reports about computer simulation.
The international character of this series is intended to be accomplished by including reports in German, English and French, both from universities and research centers in the whole world. To assure this goal, the editors' board will be composed of representatives of the different countries and areas of interest.

Interdisziplinäre Systemforschung
Analyse — Formalisierung — Simulation

Die Systemwissenschaft hat sich aus der Verbindung mehrerer Wissenschaftszweige entwickelt: der Regelungs- und Steuerungstheorie, der Kommunikationswissenschaft, der Modelltheorie und der Informatik. Sie erfüllt heute das Programm, das Norbert Wiener mit seiner Definition von Kybernetik ursprünglich vorgelegt hat und dessen Durchführung zu seiner Zeit durch die noch ungenügend entwickelte Computerwissenschaft stark eingeschränkt war.
Die Forschung und die praktische Anwendung der Systemwissenschaft bezieht heute sowohl die Fachleute der Systemwissenschaft als auch die Spezialisten der Anwendungsgebiete ein. In vielen Bereichen hat sich diese Zusammenarbeit mittlerweile bewährt.
Die Reihe «Interdisziplinäre Systemforschung» setzt sich zum Ziel, dem Studenten, dem Theoretiker und dem Praktiker über den neuesten Stand aus Lehre und Forschung, aus der Anwendung und der metatheoretischen Kritik dieser Wissenschaft zu berichten.
Dieser Rahmen soll noch insofern erweitert werden, als die Reihe in ihren Publikationen die mathematischen Modellierungsverfahren mit einbezieht, die in verschiedensten Wissenschaften in vielen Jahrzehnten zur Beschreibung und Optimierung von Systemen erarbeitet wurden.
Entgegen der früheren Tradition, in der die theoretische Regelungs- und Computerwissenschaft auf den Kreis der Mathematiker, Physiker und Ingenieure beschränkt war, liegt die Betonung dieser Reihe auf der Interdisziplinarität, die die Systemwissenschaft mittlerweile erreicht hat und weiter anstrebt. Stadt- und Regionalplaner, Psychologen, Physiologen, Betriebswirte, Volkswirtschafter, Ökologen, Ernährungswissenschafter, Soziologen, Politologen, Juristen, Pädagogen, Manager, Diplomaten, Militärwissenschafter und andere Fachleute sehen sich zunehmend mit Aufgaben der Systemforschung konfrontiert oder sogar beauftragt.
Die ISR-Reihe wird Forschungsberichte — einschliesslich Dissertationen —, Vorlesungsskripten, Readers zu Vorlesungen und Tagungsberichte enthalten. Die Verwendung wenig aufwendiger Herstellungsverfahren soll dazu dienen, die Ergebnisse der Autoren in kürzester Frist einer möglichst breiten, interessierten Öffentlichkeit zur Diskussion zu stellen. Um auch die Reproduzierbarkeit der Ergebnisse zu gewährleisten, werden in Berichten über Arbeiten mit dem Computer wenn immer möglich auch die Befehlslisten im Anhang mitgedruckt.
Der internationale Charakter der Reihe soll durch die Aufnahme von Arbeiten in Deutsch, Englisch und Französisch aus Hochschulen und Forschungszentren aus aller Welt verwirklicht werden. Dafür soll eine entsprechende Zusammensetzung des Herausgebergremiums sorgen.

ISR 67

Interdisciplinary Systems Research
Interdisziplinäre Systemforschung

Tim O'Shea

Self-Improving Teaching Systems

An Application of Artificial Intelligence to Computer Assisted Instruction

Springer Basel AG 1979

CIP-Kurztitelaufnahme der Deutschen Bibliothek

O'Shea, Tim:
Self-improving teaching systems: an application of artificial intelligence to computer assisted instruction / Tim O'Shea. — Basel, Boston, Stuttgart: Birkhäuser, 1979.
(Interdisciplinary systems research; 67)

ISBN 978-3-7643-1075-2 ISBN 978-3-0348-6553-1 (eBook)
DOI 10.1007/978-3-0348-6553-1

Originally published by Birkhäuser Verlag Basel in 1979.

ABSTRACT

This thesis is an investigation into the feasibility of constructing teaching programs which can improve their own performance. A design is presented for a self-improving teaching system comprising two principal components. One component is an adaptive teaching program where the teaching strategy is expressed as a set of production rules. The second component performs the self-improving function of the system by making experimental changes to the set of production rules. This component employs a deduction procedure which operates on a theory of instruction expressed as a set of modally qualified assertions. These assertions relate educational objectives to modifications which can be made to the teaching strategy. The cycle of operations proposed for the system is as follows - select an educational objective, make an experimental change in teaching strategy, statistically evaluate the resulting performance, and update both the set of production rules and set of assertions.

A self-improving system was implemented for the teaching of quadratic equations by the discovery method. The system was used by fifty-one students, and executed five experimental changes on its teaching strategy. This trial demonstrated that it was capable of improving its performance as a result of experimentation. Its limitations include a vulnerability to problems of local optima during 'hill-climbing' and to a variant of the frame problem. The thesis concludes by discussing the range of application of the design and outlining some further research that could be carried out in this area.

ACKNOWLEDGEMENTS

My principal debt is to my supervisor, Derek Sleeman, for the patient help, advice and encouragement he has so freely given during the last seven years. Roger Hartley and John Self have also generously spent much time helping me to clarify my ideas. I am grateful to Richard C. Anderson, Woody Bledsoe, John S. Brown, Jim Howe, Don Waterman and Jon Wexler for helpful discussions at various stages of the research.

The preliminary experiments were conducted with the help and co-operation of the Headmaster, Staff and pupils of the Hunslet Moor Junior School, Leeds. The implementation of the program was made possible by the generous support of Robert Simmons. The experiments with the program were conducted with the kind permission and active assistance of the administrative officers (in particular Mr. Schilab), the mathematics faculty (in particular Mrs. Shelton, Mrs. Waggoner and Mrs. Conway) and the students of the Austin Independent School District.

I was financially supported by an SRC award while at Leeds University, and at the University of Texas my support came from NSF Grant GJ 509X.

I thank Karen MacQueen and Jenny Rouse for typing the thesis in draft form, and Cornelia Boldyreff, Benedict du Boulay, Gordon Plotkin, Martha Stone and Richard Young for their critical comments on the draft.

I am indebted to Hoffmann La Roche for help and support during the production of the final draft, Andrew Russell for drawing the figures, and Jean Duckman for her expert and conscientious work in producing this thesis in its final form.

<u>INDEX TO FIGURES AND TABLES</u>

LIST OF CONTENTS

		Page
APPENDIX VI	Set of assertions used by the deduction system	171
APPENDIX VII	Student comments	173
APPENDIX VIII	The production rules for the pre-test administrator	175
APPENDIX IX	Glossary giving intended meanings of rare or non-standard terms	177
SUBJECT INDEX		180
AUTHOR INDEX		182

CHAPTER 1

Introduction

1.1 Guide to contents of thesis

One of the ways in which Computer Assisted Instruction (CAI) programs compare badly with human teachers is that they do not benefit from their teaching experience. A CAI program which teaches poorly in some way will teach poorly in exactly the same way after teaching another 10,000 students. This thesis is an investigation into the feasibility of constructing teaching programs which can improve their own teaching performance. In the research described here, a number of ideas and techniques from Artificial Intelligence (AI) were developed and applied to the construction of self-improving CAI programs.

This chapter sets the context of the research. Contemporary work in CAI is reviwed and illustrated by discussion of a number of representative CAI programs. The influence of work in AI on the construction of 'intelligent' CAI programs is briefly described. The various attempts in AI to construct programs with some 'learning' capability are examined for techniques and insights relevant to our goal of constructing self-improving CAI programs. The chapter ends with a more precise statement of the research objectives adopted in this thesis and some comments on the research methodology employed.

The second chapter presents a design for adaptive self-improving teaching systems. The chapter starts by considering components of CAI programs which might be amenable to automatic self-improvement. These components are then discussed with respect to the ease with which they might be automatically manipulated and the potential utility of such manipulation. The design objectives are then given by characterising an ideal self-improving CAI program. The overall structure of the design is presented. It comprises two principal components. The first is an adaptive teaching component which carries out the teaching function of the system. This component can be regarded as an automated 'teacher'. The second component incorporates the self-improving function of the system, by observing teacher performance, making explicit experimental changes in teaching strategy and evaluating such changes after teaching more students. This component can be regarded as a model (or parody) of an educational psychologist observing the automated teacher. Alternatively, it can be regarded as the self-critical function of the automated teacher.

The design of the teaching component draws heavily on the framework for adaptive teaching programs put forward by Hartley (1973). This framework comprises a vocabulary of teaching operations, a model of the student, a representation of the task and a set of means-ends guidance rules to control the teaching operations. Production rules are advocated as a suitable formalism in which to express teaching strategies. This approach is based on Waterman's (1968) work on the machine learning of heuristics which showed how strategies expressed as sets of production rules can be automatically manipulated. The chapter concludes by discussing a schema and mode of operation for the self-improving component. The schema incorporates a deductive program in the style of Black (1968) and which operates on assertions which relate experimental actions (changes executable on the set of production rules) to the overall teaching goals. The set of assertions will depend on the particular subject being taught and can be regarded as a 'theory of instruction' for that subject. The mode of operation proposed is as follows: make an experimental change in tutorial

strategy, statistically evaluate the resulting new teaching per-
2
formance and update both the tutorial strategy and the 'theory of instruction'.

The third chapter shows in detail how the techniques from AI necessary for the implementation of the design can be developed and applied to various types of CAI program. The chapter starts by illustrating how individual components of particular (real and hypothetical) CAI programs could be expressed as sets of production rules. A production rule amender is then described. The limitations of this amender are discussed both in principle and with respect to the intended practical application. The language which will be used to express theories of instruction is developed and some examples of possible theories of instruction for various teaching programs are presented. Next a deduction procedure is detailed along with examples of its use. This procedure is used to operate on theories of instruction to generate hypotheses about teaching strategies to be tested experimentally. An evaluation procedure for evaluating the results of such experiments and a generalisation procedure for adding new statements to the theory of instruction are described. The chapter concludes with a critical analysis of the deduction procedure.

The fourth chapter is concerned with a particular implementation of the design presented in the second chapter. The techniques listed in the third chapter provide the means for this implementation. The sample task chosen was that of teaching the solution of quadratic equations by the discovery method. A preliminary experiment which was carried out with twenty schoolchildren is then discussed. The analysis of this experiment was used as a basis for constructing a program which could teach the solution of quadratic equations by the discovery method. The relationship between the analysis of the preliminary experiment and the structure of the teaching program is carefully described. The program was then made into a self-improving system by adding an experimenting component. This component and its associated theory of instruction are analysed in detail. The chapter concludes with various implementation notes including the task-dependent and task-independent features of the program.

Experiments with this self-improving CAI system for the teaching of quadratic equations are the subject of the fifth chapter, which starts with a discussion of the various types of experiment which can be carried out on CAI programs. The performance of the program as a teaching program is reviewed. This was satisfactory and some notes on the program in use and the response of students to it are given. The system was first 'primed' by running it with twenty students without the self-improving component. The experimenting component was then added to the teaching program. The program generated, carried out, and evaluated five experimental changes in teaching strategy. These experiments are discussed in detail. They demonstrate that the program was capable of improving its teaching performance as a result of experimentation. The program's potential for large-scale use is then appraised in the light of these experiments.

The last chapter reviews the contributions of this thesis to CAI and to AI. The principal contribution to CAI has been to demonstrate the feasibility of constructing adaptive self-improving teaching programs and to describe the means by which such programs may be implemented. The principal contribution to AI has been to extend Waterman's (1968) work on the machine learning of heuristics and to apply it to a domain where actions are evaluated by experiment and statistical inference. A further contribution has been to illustrate how research in AI and CAI can be complementary. AI benefits when its approaches and techniques are used and tested in a complex real-world application such as teaching. CAI gains in that if AI techniques are to be applied to a class of CAI systems, then the functions, goals, and interactions of the various components of these systems must be made both explicit and as transparent as possible. The range of application of the design is indicated and its practical feasibility discussed. The chapter concludes by reviewing some of the questions raised and not answered by this thesis and by outlining further research that could profitably be carried out.

1.2 Computers in Education

Introduction

Computers are now being used extensively in Education. The profusion of acronyms -- CAI, CAL, CBE, CBL, CMI ... (where C = Computer, A = Aided or Assisted, B = Based, M = Managed, I = Instruction, L = Learning and E = Education) -- reflects the variety of ways in which they are being employed. The main classes of application are those in which

1) the student writes his own computer programs
2) the student uses the computer as an educational resource (that is, its role is analogous to that of a school laboratory or library)
3) the student is taught by the computer
4) the computer aids the teacher in his management role.

In applications of the first kind, the student may be learning a generally used programming language (for example, FORTRAN). Alternatively, he may be learning a special purpose programming language such as LOGO (Papert, 1973). In the case of LOGO, the intent is that the student will acquire general problem-solving skills as a result of writing and debugging programs in a friendly transparent language.

One example of the use of computer as educational resource is a simulation package such as the statistics program at Leeds (Hartley and Sleeman, 1971). This package functions as a simulated laboratory, the student carrying out and planning statistical experiments without having to collect data or perform tedious calculations. Another example of type (2) is an information retrieval system. Carbonell's (1970a) program SCHOLAR answers simple factual questions about the geography of South America such as "What is the population of Uruguay?".

In practice the different classes of application discussed above are not necessarily distinct. The student may have to write programs in some programming language in order to use a simulation package (for example, di Sessa's (1975) ORBIT system). A teaching program may be 'mixed-initiative' and ask questions as well as answer them (the original conception of SCHOLAR was as a teaching program). This thesis, being a study of automatically improving the quality of the teaching decisions of CAI programs, necessarily discusses CAI programs principally with respect to their teaching roles, that is, applications of type (3).

Teaching programs can be discussed on a number of dimensions including teaching style, subject matter taught, the internal representation of teaching material and response-sensitivity. The last of these is a particularly important feature of teaching programs.
To assert that a teaching program is more response-sensitive than another teaching program is to claim that in some sense it is more

adaptive to the individual learning needs of the students being taught than the other program. In order to give a brief survey of contemporary work in CAI and to clarify a number of CAI terms which will be used in later chapters, three representative pieces of work in CAI are discussed in detail below. A glossary of special CAI and AI terms used in this thesis is given in Appendix IX.

Smallwood's program

This program (Smallwood, 1962) uses an expository teaching style to impart a miniature geometry. The teaching material is organised into a number of pre-stored blocks or frames of information. After each block has been presented to the student he is asked to write down the answer to a test question. The program then shows him a number of alternative answers (as in a multiple choice test) and the student chooses the answer which most clearly matches his own. This choice determines which information block the program next branches to. The response-sensitivity of the program depends on the variety of different branches and the quality of the branching decisions. Many CAI programs take a very similar form to this program and are implemented as a branching network with pre-stored material and multiple-choice questions associated with each node of the network. From the point of view of the student such programs are not unlike automated programmed learning texts. A number of special purpose programming languages - 'author languages' - such as TUTOR, Coursewriter, etc. (Lekan, 1971) are available for implementing such CAI programs.

Smallwood's program is exceptional in that the teaching strategy is automatically changed as a result of the program's teaching experience. Smallwood's program, in fact, represents the first attempt to construct a self-improving teaching program. The improvement takes the form of 'systematic improvement in the quality of the branching decisions'. For a student at a particular level of mastery the best block to be next presented is taken to be the best block for past students who had similar histories (that is, in some sense have executed similar paths through the branching network) to that level of mastery. Smallwood uses a measure of improvement in teaching performance based on keeping the expected number of student errors below a threshold and then minimising the expected time to complete the course of instruction. One of the least satisfactory aspects of Smallwood's program is the measure of similarity of path employed. He uses the fraction of correct responses made by the student in the past. The choice of successor blocks is made by employing a bayesian model to maximise the choice of 'best' successor block. In order to evaluate untried or seldom-used paths, successor blocks are also chosen at random (with a pre-set probability).

Smallwood proposes the use of other measures of similarity (for example, reading speed or number of paths shared through the branching network) where these measures would also be used for branching decisions. The following major problems arise with Smallwood's approach:

(1) It is difficult to tell after the program has been run with students whether the teaching strategy has changed (see Smallwood (1962) p.102).

(2) As Smallwood observes, if the measure of similarity is narrow, then for a large network the program must be run with a very very large number of students for any changes in teaching strategy to occur. If the measure of similarity is broad then the program will only be capable of very limited individualisation.

(3) There is no useful task-independent measure of similarity for paths through a branching network that can be applied to this problem. The similarity of paths will depend on the teaching function and content of the material associated with the various blocks.

(4) It is not possible for the program (or its designer) to run experiments on aspects of the teaching situation other than individual branching decisions in the network and apply these results to the branching network as a whole.

These problems result from the manner in which knowledge about the student and the teaching situation is organised in the program. There is no explicit student model or theory of instruction which could be used as a basis for generalising the information obtained by experimenting with the branching decisions. The lack of such a theory also precludes the possibility of generating hypotheses about potentially useful experiments. Within this framework any attempt to greatly individualise instruction will result in a combinatorial explosion of student types and alternative paths through the network.

Smallwood (1970) has described the problem thus

> "If there are five instructional alternatives at each decision node in the tree and if there are two possible responses by the student for each instructional alternative, and if we desire to calculate the optimal instructional alternative based on these paths by the student that extend ten presentations into the future, then this will require the consideration of 10 billion possible student trajectories for each decision."

<u>Kimball's integration program</u>

This program (Kimball, 1973) operates as a tutor for integration. The student carries out integrations at a computer console and the program checks each transformation to see that it has been applied correctly. It also queries the student if he applies a transformation which is very unlikely to lead to a solution of the problem. The program's teaching style is to monitor and comment on the student's problem-solving activity. This program is an attempt to implement L. Siklòssy's (1970) slogan "computer tutors should know what they teach". It does not use pre-stored solutions to integration problems. Instead it draws on Moses' (1967) SIN, an AI program for integration, and Hearn's (1970) REDUCE, a program for algebraic simplification. Using these programs it can carry out all the integral transformations expected of the student. The program also has an archive of problems and solutions and will select examples for the student if requested. The program's estimate of the student's current skill at integration is based on the student's answers to a query about his ability to integrate and a record of the transformations the student has correctly and incorrectly applied to date. The teaching strategy is based on 'student trouble thresholds'. When one of these is exceeded the program intervenes in the student problem solution. These thresholds are based on how successful the student has been in the past, the 'likeliness' of the transformation being applied, and the expected length of the solution path. The intervention may take the form of a warning or the suggestion that the student apply a particular transformation.

Kimball's program is also self-improving in the sense that if a student solution to an archive problem is 'better' (has fewer steps) than an archive solution, then this student solution is adopted and becomes the new archive solution for that problem. The archive solutions are used in the generation of new problems and for the selection of hints. So changes to archive solutions have the effect of changing the teaching performance. Note that such changes in the tutor do not affect the response-sensitivity of the integration tutor. The program is also restricted to its initial set of archived problems for teaching material and is vulnerable to the acquisition of spurious 'better' student solutions. For example, a condensed sequence of transformations only applicable to a particular problem may be copied from a student by the tutor. Kimball was in practice forced to adapt his program to trap single-step 'guess' solutions but his program remains vulnerable to longer spurious solutions.

Leeds Arithmetic

A series of arithmetic teaching programs have been implemented at Leeds (Woods et al., 1969; Woods and Hartley, 1971; Tait et al., 1973). These programs generate and administer drill and practice arithmetic problems of various levels of difficulty. The response-sensitivity of these programs stems from their ability to (1) generate material suitable to the student's level of competence, (2) administer different types of feedback, and (3) generate remedial material depending on student errors. The programs are structured according to Hartley's (1973) framework for adaptive teaching programs. According to this framework, adaptive teaching programs are viewed as comprising four components:- a representation of the task; a representation of the student (student model) and his performance (student record); a vocabulary of teaching operation; and a pay-off matrix, or set of means-end guidance rules. In a well-documented series of experiments (Woods and Hartley, 1971) a model of task difficulty based on number of rows and digit size was developed and validated. The student model includes the student's actual working level of success and the error factor estimates associated with the operations the student is using in calculations. One aspect of the teaching strategy centres on generating examples of such difficulty that the student operates at a particular level of success. For addition, experiments were carried out to determine the effect on various classes of students of working at different levels of success. For subtraction, experiments were carried out with two alternative task difficulty models (corresponding to 'counting up' and 'counting down') and it was discovered that while half the subjects used the former strategy and some the latter, there were a considerable number of students who were versatile and who employed the method most convenient for the particular problem they were solving. In multiplication, experiments were carried out in which different types of feedback (that is, location of error and remedial isntruction) were presented to the student. One group of students had no feedback (except number of correct answers), one group of students was given 'passive' feedback (knowledge of correct answers) and one group was given 'active' feedback (on making a mistake the student was engaged in a 'dialogue' by the program and the steps in the problem were explained). Students with feedback performed significantly better than those without feedback and there was some evidence that weaker students benefited more from the active feedback than from passive feedback.

The development of the Leeds arithmetic program was the main CAI influence on the work described in this thesis. The experiments and development were based on explicit task analyses (the task difficulty models), explicit measures of student performance (rate of working, actual level of success, error-types, pre- and post-tests), and a number of hypotheses related to changes in teaching strategy (based on level of success of working, competing task difficulty models, and type of feedback). There seemed to be no reason, in principle, why experiments and refinements in teaching strategy, such as those described above, could not be carried out automatically by the teaching program itself. The fundamental problem was to find some suitably explicit form for expressing the task analysis, student model, and teaching strategy so that they would be amenable to automatic manipulation.

Summary

The three programs described above are very advanced pieces of CAI and each one taken as a whole is quite atypical of the state of the art of CAI. Very few self-improving CAI programs have ever been implemented and, unlike the Leeds arithmetic programs, the great majority of CAI programs are implemented without reference to any educational or psychological experiments on the subject matter to be taught. However, between them the three programs do employ most of the techniques used in the construction of computer tutors. The teaching styles described are variously (a) expository (Smallwood), (b) tutorial or problem-solving monitor (Kimball), and (c) drill and practice (Leeds). The internal representations of the task are respectively (a) branching network (with pre-stored frames), (b) an algorithm to perform integrations, and (c) a task difficulty model (which is employed to generate examples). The response-sensitivity is related to a student model which operates on (a) the last answer to the last question at the last node, (b) the student's answers to a pre-test quiz, and (c) the student's level of success and error types.

Now, compared to human tutors, most CAI programs which have an active teaching role are severely limited. The limitations include

(1) inability to conduct dialogues with the student in natural language

(2) inability to understand the subject being taught, in the sense that the program cannot accept unanticipated responses

(3) inability to understand the nature of the student's mistakes or misconceptions

(4) inability to profit from experience with students or to experiment with the teaching strategy.

These limitations result in part from the way teaching and subject matter knowledge are represented and organised in CAI programs. As such, these shortcomings have been tackled as AI (or 'AI in CAI') problems by a number of workers (see, for example, Carbonell, 1970b; Pask, 1972; and Koffman, 1973). Perhaps the most successful CAI program with regard to problems (1) and (2) is Brown et al.'s (1976) SOPHIE. This program teaches electronic trouble shooting and employs a variety of alternative representations of the skill being taught for use in both interpreting the English input and also following the student's reasoning. Other examples of programs that 'know what they teach' are Goldberg's (1973) program, which teaches theorem-proving (and incorporates a heuristically programmed theorem-prover), and Sleeman's (1974) problem-solving monitor for NMR spectroscopy. No really substantial contributions have been made to problem (3). The advances which have been achieved result from work in Educational Psychology rather than AI. Such advances have been derived from experiments in particular teaching domains such as the various arithmetic programs (Woods and Hartley, 1971; Suppes, 1971; Howe and Cassels, 1974). Both Self (1974) and Stansfield (1974) have proposed that student models should be both explicit and represented as procedures. However, this application of the AI procedural embedding of knowledge thesis (see Hewitt, 1972) has not been successfully applied to a working CAI program. For the most part the student models implicit in contemporary CAI programs are rudimentary.

The immediate focus of this thesis is problem (4) -- the lack of learning capability in CAI programs. As an AI approach is adopted, the main issue is naturally one of representation. To my knowledge only two CAI programs reported in the literature have a self-improving capability -- that of Smallwood and that of Kimball. The approach taken here is radically different from their work. In Smallwood's program, random local changes in branching decisions are made and evaluated. The teaching strategy, task analysis, and student model are all implicit, and it is not possible to carry out and record the result of experiment on aspects of these components. In Kimball's program, sequences of integration transforms shorter than those which would have been carried out by the program/tutor are learned from the student. So the program's ability to carry out integrations improves and hence its hints to the students may change, but its response-sensitivity as such (which depends on the 'student trouble thresholds') remains fixed.

In this thesis the aim was to find a way to express teaching strategies so that it would be possible to 'prime' the system with general and specific hypotheses about alternative ways of teaching. The system should then carry out experiments to test these hypotheses and amend the teaching strategy accordingly. The model for this experimental activity was the evolution of the Leeds arithmetic programs.

1.3 Programs that learn

In this section we discuss a number of AI programs with learning capability. This discussion focusses on the potential applicability of the techniques employed to our objective. This is the construction of self-improving teaching programs, so we require a descriptive system in which the following are possible:

i) The formulation of teaching programs.

ii) The modification of the teaching strategies of such teaching programs.

iii) The permanent incorporation of such modifications as the result of experiments on teaching strategies.

iv) The development of such experiments from hypotheses about the teaching strategy being employed.

v) The generation of such hypotheses from a theory of instruction.

The terms 'formulation', 'modification', 'result', 'development', and 'generation', in (i) to (v) respectively, are being employed in both a weak and a strong sense. The weak sense relates to the practical ease of reprogramming the teaching program and analysing its performance. The strong sense corresponds to the program changing itself (for (ii)) and generating its own hypotheses (for (v)).

In practice, the descriptive systems in which most learning programs are expressed are high-level programming languages. So, as a question of theoretical principle, what can be expressed in one such system will be expressible in most other such systems. The notion of possibility referred to with respect to (i) to (v) is a practical or pragmatic question. Below we give a review of some of the attempts to construct programs that learn. In each case we discuss how easily (i) to (v) above could be achieved by employing the descriptive system used in the construction of this particular learning program. This brief review is confined to exemplars of the approaches which were actually considered in the design of the self-improving program. A more comprehensive review of the literature is given in Jones (1971). Some other relevant and more recent work is discussed in Chapter 6.

Perceptrons

The perceptron is a good example of a learning program which learns by quantitative adaptation. In its simplest form it carries out pattern recognition by means of a weighted average of its inputs (see Rosenblatt, 1957). It is trained to discriminate patterns by being given examples and counter-examples of the class of patterns to be recognised. For a single-layer perceptron it is possible to prove a Perceptron Convergence Theorem (see Minsky and Papert, 1969). This states that if weights exist such that a perceptron could discriminate a class of patterns, then these weights can be found in a finite number of training steps. This result seems very impressive, but Minsky and Papert in their analysis of the capabilities of single-layer perceptrons have proved that sets of weights cannot exist for many important topological concepts, in particular 'connectivity'. Minsky and Papert have also conjectured that similar limitations apply to multi-layered perceptrons. However, multi-layered perceptrons have not to date proved tractable to mathematical analysis and it is much harder to prove convergence theorems for them or to establish their limitations (see Block, 1970).

Smallwood's teaching program is in some ways analogous in structure to a multi-layer perceptron with very few inputs to each component simple perceptron (although the way in which inputs are combined is considerably more complex than in simple perceptrons).

As Smallwood has demonstrated, it is certainly possible to achieve objectives (i) and (ii) for teaching material organised in pre-stored frames in such a structure. The obvious limitation is that all the experiments carried out are (as in the perceptron) very local in character, and the results of any such experiment are reflected in changes in weights. The teaching strategy and the results of these 'local experiments' are implicit, with the attendant disadvantages discussed above in 1.2.

Samuel's Checkers Program

This program, implemented by Samuel (1959, 1960, 1967) improves its performance by quantitative adaptation based on its game-playing experience. The program plays using a game tree in which 'look-ahead' minimaxing techniques (see Nilsson, 1971) are employed in conjunction with a linear polynomial which is used to evaluate board positions. This polynomial is a weighted sum of board features. The program's learning takes the following three forms. Firstly, storing board positions which occur during games with an associated measure of their utility. Secondly, modification of the weights (coefficients of the scoring polynomial) as a result of play. Thirdly, training consisting of board positions and the best plays associated with the positions, the training information being used to modify the weights. The first of these is a form of rote learning; the other two are forms of generalisation. Samuel's program has been successful both as a checkers-player and as a program that learns to improve its game-playing performance.

Samuel's approach could be applied to the construction of a self-improving computer tutor in the following manner. Let us view the teaching operations for some domain as moves in a board game. The component of the computer tutor used for the generation of teaching material would then correspond to the move selector. If the student model could be expressed as a linear polynomial then predicting the student's current state of knowledge would correspond to evaluating board positions. This could be used by a computer tutor similar to Smallwood's which generated (as opposed to selected prestored) teaching material. So the result would be a generative teaching program with a self-improving capability with objectives (i) to (iii) for self-improving teaching programs being met. The objections to this scheme are various. There is the general objection which also applies to perceptrons that much of the knowledge is represented implicitly. Hence, although this scheme might 'work' for generative teaching programs, all the major criticisms made of Smallwood's structure would still hold. In particular, objectives (iv) and (v) would be just as difficult to achieve. Further, it would be necessary to express student models as linear polynomials of independent variables. This represents a very severe restriction -- particularly in view of the well-argued case (see Self, 1974) for expressing student models as collections of procedures.

In the teaching situation, many of the 'teaching moves' have different functions (some will be related to overall teaching goals, some related to subgoals, some will be used to elicit more information about the student's state of knowledge, etc.) and many of these moves will be closely related. To 'learn to teach better' effectively we will need to express these functions and interrelationships. Any scheme such as Samuel's based on 'quantitative adaptation' which employs 'homogeneous and unstructured arrays of numbers' (Minsky and Papert, 1972) does not facilitate this. Similar objections apply, of course, in many other learning domains, as well as in the teaching domain.

Learning structural descriptions from examples

Winston's (1970) program forms object definitions such as 'arch'. The program operates in the MIT toy 'blocks' world and is trained by being given examples and counter-examples of particular concepts. It operates by building and modifying descriptions in terms of networks in which nodes represent objects and arcs represent relations. It is also possible to have arcs connected to other arcs (that is, relations between relations). The program is capable of acquiring simple concepts and then applying them to other situations. For example, in his thesis, Winston claims that having acquired the concept 'arch' the program can then identify a row of three arches as such (rather than nine blocks in a very complex relationship).

Winston's work adds to our understanding of how to describe the decomposition of visual scenes. However, the essential point is that the learning scheme he proposes is not 'robust'. It depends crucially on sequences of 'near-misses' -- which must differ in one (and only one) attribute from the new concept currently being acquired. On even a superficial examination of most teaching situations, it becomes clear that in order to learn about the varying and idiosyncratic learning behaviour of students we must employ very robust learning schemes. The behaviour of an educational researcher in the classroom is hardly similar to that of a subject in a concept-identification experiment (the behaviour of Winston's program is, however, closely akin to that of an ideal subject in such an experiment (see Young, 1976)).

Despite these comments and various contemporary criticisms of the learning mechanism proposed by Winston (Banerji, 1974; Knapman, 1977), his work is of crucial importance. This is because by employing knowledge related to the task domain in an appropriate form of description, Winston succeeded in constructing a learning program with an 'understanding' of the task domain in which it was operating. So his work is the first clear demonstration of the thesis that the central issue in the construction of learning programs is the development of appropriate descriptive systems.

Machine learning of heuristics

Waterman (1968, 1970) has considered the problem of strategy learning in game-playing and problem-solving environments. He has shown how production rules (Post, 1943) can be used to represent heuristics in an explicit form amenable to automatic manipulation. He has demonstrated how it is possible to evaluate heuristics expressed in this form and modify them as a result of either training or operating in the task environment. Waterman successfully employed these techniques in a series of programs which played poker and improved their betting decisions. The version of this program which was not explicitly trained used an axiomatisation embodying a 'theory of poker' to evaluate its betting decisions.

Waterman has thus explicated and demonstrated a method for describing strategies of action which facilitate machine learning. In fact, a system of production rules can be viewed as a programming language in which the only permissible statements are certain types of conditional. Waterman's work suggests a solution to the central problem in the construction of a self-improving teaching system, namely a suitable formalism for the expression of the tutorial strategy and the student model.

Note also that poker is not a game of perfect information (see Gale, 1960). In poker a player does not have access to all the information as he cannot see his opponent's hand and may never do so if his opponent discards his cards. Accordingly it is not possible to construct a game-tree and employ some form of look-ahead and minimaxing procedure to discover the 'optimal' move. Instead it is necessary to estimate the probability that your opponent is bluffing. There is a close analogy here with the teaching context in which we are continually estimating the student's current state of knowledge.

After reading Waterman's thesis it became clear to me that the construction of a self-improving teaching system which met objectives (i) - (v) was feasible. There are, of course, a number of ways in which the teaching context is very different to the game-playing one, and the following questions arise in attempting to apply to the teaching domain a scheme similar to Waterman's.

A. How exactly can teaching strategies be expressed as sets of production rules?

B. How can teaching experiments be carried out and evaluated in a similar fashion to that employed by educational researchers in the development and testing of teaching programs?

C. How can a theory of instruction for a particular teaching domain be axiomatised?

These questions are discussed both in general and specific terms in subsequent chapters.

1.4 Research Objectives and Methodology

The primary research objective of this thesis is to examine the feasibility of automating the process of improving the quality of instruction of CAI programs. For the purposes of this thesis we consider only teaching situations for which the following assumptions can be made:

1) The set of possible teaching actions can be made explicit.

2) A set of explicit teaching objectives can be given.

3) The progress of individual students towards meeting these teaching objectives can be measured.

4) A set of general educational objectives can be associated with the teaching situation as a whole.

5) Measures can be given which determine progress towards these general objectives.

For the implementation of any response-sensitive (or adaptive) teaching program, it is necessary to be able to make assumptions (1), (2) and (3). If assumptions (4) and (5) can be made, then we will define 'an improvement in the quality of instruction' as a change in the teaching strategy which results in greater achievement of the general educational objectives of the teaching situation. To make this notion precise, it will be necessary to define an overall educational goal, and this can be very difficult. Often there will be a number of conflicting goals associated with a teaching situation. In this case we must assume that it is possible to associate relative weights to the educational objectives and that an overall goal can be defined in terms of some function of these objectives and their weights. Defining such educational objectives, while very difficult, can be of considerable value in forcing a more precise and careful analysis of the teaching situation.

Note that in this thesis the various teaching situations are taken as fixed and are then analysed and an attempt made to improve the quality of the teaching with respect to some set educational objectives. Much innovative work in education is concerned with attempts to promote new educational objectives (see, for example, Leonard, 1968; Postman and Weingartner, 1969; Papert, 1973). It is clear that for many domains major reformulations will be necessary to achieve any substantial improvement in the quality of education. Such questions are outside the scope of this thesis, which is restricted in application to the domains and styles of teaching employed in contemporary CAI programs.

There is an important assumption implicit in any attempt to continue indefinitely improving the quality of instruction. This assumption is that the models of learning employed in CAI programs will usually be inadequate (except in trivial cases) in that they will not always correctly predict, directly or in a probabilistic way, the change in state of a student's knowledge which results from the execution of the teaching action. Even a brief perusal of the CAI learning theory literature will show that this is not a contentious assumption -- in Atkinson's (1972) words "the models of learning that now exist are totally inadequate to explain the subtle way by which the human organism stores, processes, and retrieves information".

In order to achieve the primary objective stated above, the following secondary objectives were adopted:

a) To develop a general design for self-improving teaching programs for which assumptions (1) - (5) are true.

b) To develop and evaluate AI techniques to make possible and facilitate the implementation of the design for particular teaching tasks.

c) To implement a self-improving teaching program for a particular domain.

d) To experiment with and evaluate such a self-improving teaching program.

The attempts to meet the secondary objectives (a) - (d) are described in chapters 2 - 5 respectively. These secondary objectives reveal our research methodology. This methodology is not uncommon in work in AI, and centres on the implementation of a sequence of working performance programs.

At each stage in the research, computer programs were implemented, debugged, and often discarded. The implementation of each of these programs was regarded as an exploration. In the course of this research three sorts of exploratory implementation were employed: firstly, those which test the adequacy of the knowledge incorporated in the program (for example, the adequacy of a student model in predicting a student's state of knowledge); secondly, those which test the ease with which a certain type of representation can be employed in the expression of a particular type of knowledge (for example, the adequacy of production rules as a programming language for expressing tutorial strategies); and thirdly, those which test how well certain programs with different functions interact (for example, how the experimenting program changes the teaching program).

The discussion in the subsequent chapters is based on experiences with these exploratory implementations.

CHAPTER 2

A design for self-improving teaching systems

2.1 Preamble

Before formulating a design for a self-improving teaching system, it is necessary to consider exactly what types of changes can be executed on a computer program with a teaching role. Some types of changes, for example changing the teaching style of the program or the subject matter taught, are possible in principle, but in practice to make this possible it would be necessary to have a collection of entirely distinct CAI programs. In this context the design of a self-improving program is trivial and would consist essentially of a 'switch' and a number of alternative teaching programs.

Another class of changes are those which operate on a teaching program's internal representation of its teaching material. In this context self-improvement would take the form of learning new facts, concepts or skills from the student being taught. For the most part such learning is completely beyond the state of the art because of the way knowledge about the teaching domain is represented in such programs. Even for those programs for which such changes might be practical there are extreme difficulties. Consider, for example, Carbonell's 1970a SCHOLAR program which teaches facts about the geography of South America by employing a semantic network. This program could, in principle, acquire new facts by changing the nodes in the semantic network. But the program would not be able to detect false facts nor resolve contradictions. The difficulties of incorporating new information into such semantic representations are discussed in Hendrix et al. (1973). Stansfield (1974) discusses the difficulty of handling contradictions in this context.

The only example of a CAI program which might be said to improve the internal representation of its teaching material is that of Kimball (1973) discussed above. The program can acquire new solutions to problems already in its archive of standard problems. Even for this very restricted type of self-improvement it is an essential prerequisite that the program have a 'knowledge' of the subject area which can be directly related to student performance. Few such programs exist at present.

The design presented in this chapter (discussed further in O'Shea and Sleeman, 1973) is directed at improving the quality of instruction by altering and experimenting with those features of CAI programs which relate to their response-sensitivity. Some specific examples of such features and how they may be changed are:-

a) The best order to present a set of concepts to students who have different learning styles. A good example of this is given by Pask and Scott (1972) in their discussion of two classes of learner distinguished on the basis of their cognitive style as 'serialists' and 'holists' by their performance in learning a taxonomy of Martian fauna.

b) The point at which a program monitoring a student performing a problem-solving task should intervene. For example, Kimball's program has a technique intervention threshold, a problem length threshold and a problem intervention threshold. When any of these thresholds are exceeded the program intervenes and performs some remedial operation. The optimal values for these thresholds are not readily apparent. Further, there are classes of students for which the value of the thresholds should be different. A competent student should probably be allowed more leeway to try out unusual approaches than a naive student.

c) The level of success at which students operate. Woods and Hartley (1971) show how the Leeds Arithmetic program can generate examples using a task difficulty model and the predicted competence of the student, so that the student will operate at a specified probability of success. As Woods and Hartley observe, "learning rates and motivation vary with probability of success for children of different personality characteristics". Accordingly, it would be of use to identify the optimal probability of success for different classes of students.

The above examples were of course chosen for the clarity with which they illustrate features of a CAI program which could possibly be changed with profit. In many educational contexts things are considerably more complex. A single piece of instruction may be used for a variety of pedagogical motives (see, for example, pp. 3-5 in Polya, 1973). For example, a single proof in geometry may be used to:-

test whether a student has mastered a particular concept,

increase the student's familiarity with a certain standard form of proof,

improve the student's general problem-solving ability, or

keep the student occupied.

2.2 Design Objectives

The expression 'teaching strategy' is often used to include such features of a teaching program as its teaching style and degree of learner control. The term 'tutorial strategy' will be used here to denote those features of CAI programs which affect response-sensitivity (for example, task difficulty models, student models, performance thresholds, presentation sequences). An *ideal* CAI program with a teaching role and the capability of improving its tutorial strategy should have the following properties:-

i) Comparable teaching performance to non-self-improving teaching programs for the same subject area.

ii) The ability to use a theory of instruction as a basis for generating hypotheses about potentially useful modifications to the tutorial strategy.

iii) The ability to evaluate such hypotheses by running experiments on the results of such modifications and the ability to refine the theory of instruction accordingly.

iv) The ability to change the theory of instruction as a result of advice.

v) Teaching performance which is not subject to sudden severe deteriorations.

vi) Teaching performance which improves as a result of teaching numbers of students.

vii) A flexible, transparent control structure amenable to change both automatically and manually.

The last characteristic is almost certainly a prerequisite in any attempt to achieve the others. There is a trade-off problem inherent in running experiments (iii) and maintaining stable teaching performance (v). The advice (iv) will have to be expressed in some formal language. Formulating theories of instruction presents formidable problems. Representing such theories in a way that they facilitate the various functions required (ii, iii, iv) presents further substantial problems.

In chapter 5 we evaluate the self-improving program implemented in the course of this research with respect to these ideal characteristics. In the rest of this chapter we present a general design for a self-improving program with these properties.

2.3 Structure

It is necessary to isolate the various processes that would be carried out by a self-improving teaching system. The design comprises a hierarchy of three separate interacting levels, where the possible interactions between them are well defined. This is illustrated in figure 1. The functions of the three levels are, respectively, to act as an interface between the student and program, to execute the tutorial strategy, and to change and run experiments on the tutorial strategy.

The first two components are equivalent to an ordinary CAI program. Let us further isolate the processes implicit in such a program by using Hartley's (1973) framework for an adaptive (response-sensitive) teaching program. This framework comprises:-

i) A vocabulary of teaching operations.
ii) A representation of the task.
iii) A model of the student.
iv) A set of means-end guidance rules.

The vocabulary of teaching operations will be just the various particular pieces of teaching material (information, exercises, questions, etc.) that can be presented to the student.

A representation of a task can, and usually will, comprise two components. Firstly, what we have termed 'the representation of task material'; this could be, for example, an algorithm for generating examples or a semantic network. The other component of the task representation may be some measure of difficulty or ordering of concepts or skills to be acquired. This, and the student model, can be

Figure 1: Outline design

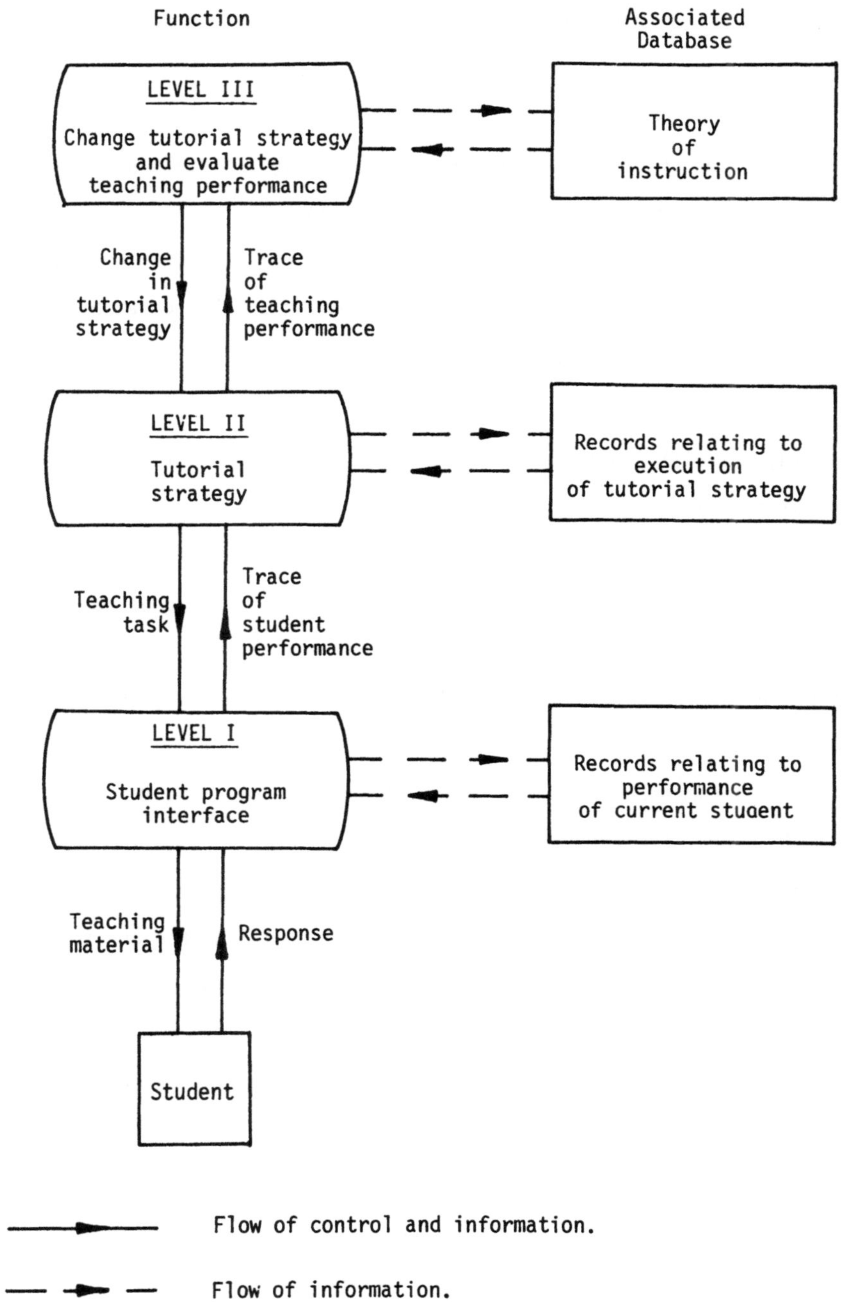

used to achieve response-sensitivity. Means-end guidance rules are used to select or generate teaching operations by providing a mapping from the state of student knowledge (as predicted by the student model) to appropriate task material features (as given by the task analysis). The means-end guidance rules will be used to express such aspects of the tutorial strategy as tutorial subgoals, termination criteria, and intervention thresholds. Together the means-end guidance rules, student model, and task analysis constitute the tutorial strategy. Figure 2 illustrates the operation of a CAI program broken down into the above components.

The function of the student program interface is to present teaching material, administer questions, or set problems to the student and maintain a record of interactions with the student. The range of possible interactions with the student depends on the structure and content of this interface. As the design presented here does not lend itself to automatically changing this component, there is a limit on the potential for change in teaching performance; that is, the repertoire of instructional operations is taken to be fixed.

This component will be a software package operating via some sort of interactive terminal; for example, teletype, CRT with light pen, or speech recognition device. In a teaching program without severe restrictions on input it could be a natural language question-answering program (such as Burton's (1976) program used in Brown et al.'s (1976) SOPHIE).

The various processes necessary for the operation of the top experimenting level can be isolated and expressed in terms of the following functions:-

i) A deduction schema or possibilities list to be used for the generation of hypotheses concerning potentially useful changes in tutorial strategy.

Figure 2: Schematic diagram of operation of an adaptive teaching program.

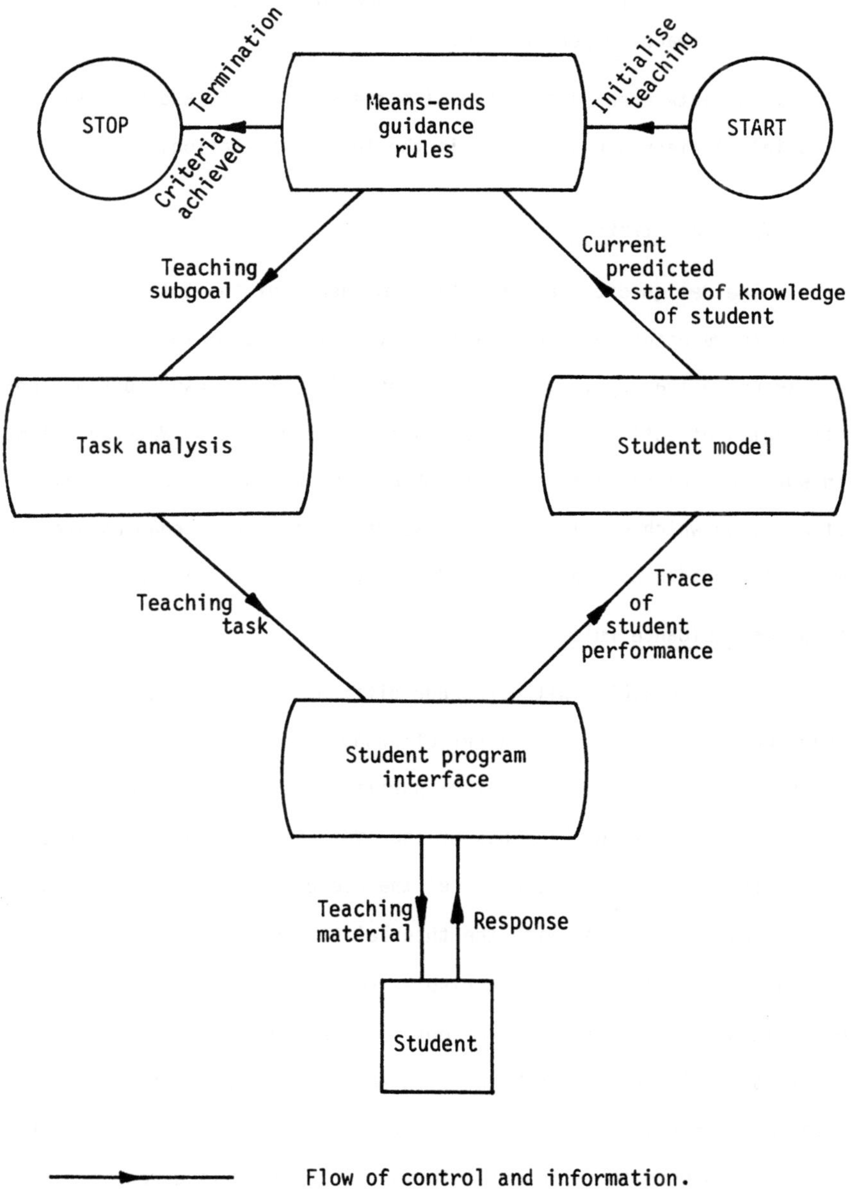

ii) An amender to execute changes on the tutorial strategy.

iii) A statistics package to evaluate the effects of such changes.

iv) A monitor to check for severe deteriorations in teaching performance.

The data bases associated with these functions and with the tutorial strategy are discussed in the following sections.

2.4 Tutorial strategy

The design dictates that the components of the tutorial strategy be represented in a form amenable to automatic manipulation. In 1.3 we noted that systems of production rules may be used to represent strategies of action, and had successfully been used by Waterman (1968) in work on machine learning. The three components of the tutorial strategy on which we will wish to execute experimental changes are the means-end guidance rules, the student model, and the task analysis.

Means-end guidance rules

It is possible, without undue difficulty, to express a set of means-end guidance rules as a set of production rules where this set comprises a state-vector, a set of partitions defined on the set of possible values of the elements of the state-vector, and an ordered set of condition-action rules. Let the elements of the state-vector be variables associated with the tutorial strategy, and let the partitions give measures on the variables relevant to the tutorial strategy. Then, if we define the various tutorial subgoals as actions, the condition-action rules can be used to relate tutorial contexts to tutorial subgoals. A detailed account, with examples of how production rules may be used to represent tutorial strategies, is given in 3.1.

Student models

A student model (see Self, 1974) is that component of a teaching program which is used to predict the current state of knowledge of the student. It usually is a simple algorithm operating on a database which includes a record of the student's past responses. For a teaching program with an expository teaching style a typical prediction might be that a student will always correctly answer some particular question. This prediction would be based on the student's past responses to some subset of questions previously asked by the program. For a drill and practice program a typical prediction might be an estimate of the probability that the student will correctly apply some skill or successfully complete some operation. Note that in any long sequence of teaching operations a problem arises as to how much of the sequence ought to be used in the estimation of such a probability (see Woods and Hartley, 1971).

An alternative approach would be to use procedures which directly simulated the learning processes of the student. This type of model (not used to my knowledge in any operational CAI program) would consist of a set of algorithms for executing the task being learned in ways that corresponded to different states of student knowledge. Prediction would involve identifying the algorithm whose performance gives the closest approximation to that of the student being modelled.

The first class of student model can easily be translated into sets of productions in an analogous fashion to that employed for the means-ends guidance rules. The actions would be particular states of student knowledge, or in the case where the predictions were probabilistic estimates they would be a function which executed some calculation on the elements of the state-vector.

The facility with which simulation models might be translated into sets of production rules is illustrated by the work of Waterman and Newell (1973) on cryptarithmetic, Baylor and Gascon (1974) on weight seriation, and Young on length seriation (1974) and subtraction (1977). Examples of the use of production rules to express student models are given below (see 3.1 and 4.3).

Task analyses

Where a task analysis consists of an ordering or partial ordering of a set of concepts, it may be translated into a set of productions in the same way as a tutorial strategy of ordered or partially ordered subgoals may be so expressed (see 3.1). A task difficulty model, of the type employed in the Leeds Arithmetic program, relates teaching goals to features of teaching material. As such it may be more amenable to representation as a matrix. It is possible to re-write such a matrix as a set of productions. However, such a matrix can be viewed and manipulated as a subset of the class of productions employed here.

2.5 Experimenting with tutorial strategies

Generating hypotheses

Any attempt to generate hypotheses about possibly useful changes in tutorial strategy (or strategies of action in general) raises very difficult and interesting problems. In our context the following are a prerequisite to such attempts:-

a) An overall educational objective explicitly defining the notion of 'improvement' in terms of measures on teaching performance.

b) A formal language in which assertions may be made, relating features (both general and specific) of tutorial strategy to the tutorial goals or subgoals associated with the overall educational objective.

c) A theory of instruction which can be expressed in the formal language of (b); ('theory' is being used in a very weak sense here).

However, defining an appropriate formal language and identifying and expressing in such a language the wide range of general, specific and common-sense knowledge implicit in the theories of instruction employed by educational psychologists is fraught with difficulties. One approach based on the deductive system of Black (1968), and its limitations, are given in the next chapter.

Amending tutorial strategies and evaluating teaching performance

Amending a tutorial strategy expressed as a set of production rules can be facilitated by using task-dependent information. An amender constructed on these principles is given later (3.2). Having changed the tutorial strategy, it is necessary to check whether the new teaching strategy is 'acceptable'; that is, to ensure that no unwanted side-effects have been introduced into the teaching performance. Examples of such constraints might be the minimum number of interactions permissible and the criteria for the successful termination of the teaching session.

One way of ensuring that these constraints are met is by continuously monitoring teaching performance. Alternatively, if the student model used in the tutorial strategy were of the simulation type, then it could be used as an 'artificial student' to test out the amended tutorial strategy. This process should detect grossly unacceptable tutorial strategies, but continual monitoring would still of course be necessary: (if the artificial student were perfect an optimal tutorial strategy could be devised, and the self-improving component would be unnecessary).

Standard statistical methods can be applied to the evaluation of teaching performance provided the teaching goals have measures associated with them. There is also the added advantage that, as the experiment is being carried out on-line, it is possible to terminate it as soon as a significant result is obtained.

Running experiments/cycle of operations

The cycle of operations of the experimenting component is: generate a hypothesis, execute a change, monitor and evaluate subsequent teaching performance, update the set of assertions which embody the theory of instruction, etc. This is illustrated in figure 3. Note that updating the set of assertions ensures that identical experiments are not repeated and that assertions leading to false hypotheses can be amended.

No precise experimental procedure is proposed here. We have not laid down the experimental design(s) to be followed, nor the type of statistical test(s) to be employed. These will vary with the types of change in tutorial strategy being carried out, the specificity of the theory of instruction, and the number and range of ability of the students (experimental subjects) available for the program to teach. Any attempt to express as a computer program one of the more complex experimental methodologies employed in practice (such as that implicit in the development and evaluation of the Leeds Arithmetic) would be a major piece of research on its own. For the purposes of this thesis a particular simple experimental methodology is developed in 3.4. The design as presented here could be used in conjunction with other experimental methodologies provided that a suitably precise theory of instruction could be given.

Figure 3: Cycle of operations of experimenting component

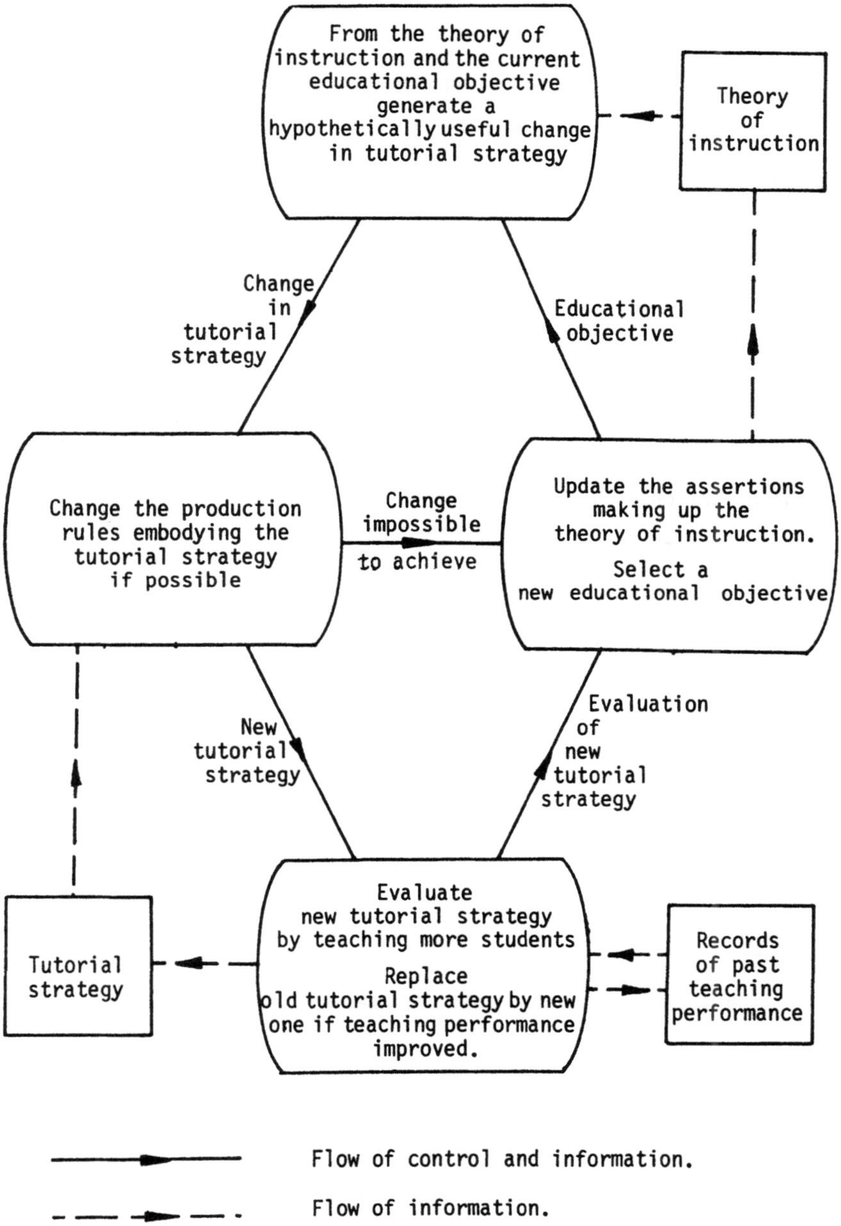

CHAPTER 3

Representing tutorial strategies and theories of instruction

This chapter describes in detail an approach to the two complementary problems raised in the previous chapter. First, how can tutorial strategies be represented so that they are amenable to automatic manipulation. Second, how can theories of instruction be represented so that they can be used as a basis for running experiments on features of such tutorial strategies.

3.1 Expressing tutorial strategies

The suitability of production rules as a way of expressing tutorial strategies in the context of self-improving teaching systems has been mentioned above (1.4, 2.3). The relevant features of systems of production rules are:

a) They can be used to represent strategies of action.
b) They provide a clear, simple, transparent structure.
c) They are amenable to automatic manipulation.

This section contains a discussion of (a) and (b) where the strategy of action to be expressed is a tutorial strategy, or a component (e.g. student model) of a tutorial strategy. The following section (3.2) discusses (c) in the context of amending tutorial strategies.

An example

The type of production systems employed in this thesis are based directly on Waterman's work, and a comprehensive discussion of them can be found in his thesis (1968). Other types of production system are reviewed by Davis and King (1975), Rychener (1976) and Waterman (1977). In the production rule systems employed here the mechanism used for conflict resolution (that is, which rule gets fired if more than one is potentially applicable) is rule order.

The use of production rules in a CAI context will first be illustrated by their application in a simple hypothetical CAI teaching situation. Consider three teaching operations T(i) (i = 1, 2 or 3), the execution of which results in the student acquiring an associated concept C(i) (with probability <1) and the testing of the student's new knowledge state. Let the acquisition of C(3) be the objective of the teaching task, and let the acquisition of C(1) or C(2) be usually, but not always, a prerequisite for acquiring C(3). Assume there is no forgetting. Suppose that with each student is associated a relevant score PS on some pre-test ($0 \leq PS \leq 10$). Further suppose it is known that, for students with high PS scores, they are slightly more likely to acquire C(1) via T(1) than C(2) via T(2). However, students with low PS scores are much more likely to acquire C(2) via T(2) than C(1) via T(1). Further, suppose that there is for some practical reason a limit of ten teaching operations per student.

Our systems of production rules comprise a state-vector, a set of partitions, and an *ordered* set of condition-action rules. For the above example a suitable state-vector would be (LASTOP, STATE, SCORE, NUMOP) where LASTOP is the last teaching operation which has been executed, STATE is the current state of student knowledge (i.e., the concepts he has acquired), SCORE is PS , the pre-test score, and NUMOP is the total number of teaching operations performed.

A suitable set of partitions might be:

```
LASTOP → LASTOP1 if  LASTOP = T(1)
       → LASTOP2 if  LASTOP = T(2)
STATE  → STATE1  if  STATE = C(1) or STATE ≐ C(2)
       → STATE2  if  STATE = C(3)
```

SCORE → SCORE1 if SCORE ⩽ 5
→ SCORE2 if SCORE > 5
NUMOP → NUMOP1 if NUMOP > 4
→ NUMOP2 if NUMOP > 7 NUMOP < 10
→ NUMOP3 if NUMOP = 10

The partitions are employed as follows. If the values of the variables in the state-vector are (T(2), C(1), 6, 8) this is parsed with the partitions to yield the symbolic state-vector (LASTOP2, STATE1, SCORE2, (NUMOP1, NUMOP2)).

An appropriate set of condition-action rules might be:

1 (*, STATE2, *, *) → STOP
2 (*, *, *, NUMOP3) → STOP
3 (*, STATE1, *, *) → T(3)
4 (LASTOP1, *, SCORE1, NUMOP1) → T(2)
5 (LASTOP2, *, SCORE2, NUMOP2) → T(1)
6 (*, *, SCORE1, *) → T(1)
7 (*, *, SCORE2, *) → T(2)

The symbolic state-vector is compared to each of the left-hand sides in turn until the first match is found -- a match occurring when all the symbols in the condition part of the rule, excepting ' * 's, are present in the symbolic state-vector. In the example given, (LASTOP2, STATE1, SCORE2, (NUMOP1, NUMOP2)) matches the third rule (it would match the fifth if the third were omitted), and the right-hand side, namely T(3), is executed and the state-vector updated accordingly.

The above example illustrates the power of this form of representation. Expressed in words, the tutorial strategy embodied in the above seven condition-action rules is as follows: "Continue teaching until ten teaching operations have been executed unless the student reaches knowledge state C(3) first in which case stop at once. As soon as a student acquires either state C(I) or C(2) start executing teaching operation T(3). For a student with a low pre-test score, make up to five attempts to impart C(1) via T(1) and then switch to execution of T(2) . For a student with a high pre-test score, make up to eight attempts to impart C(2) via T(2) and then switch to T(1) ."

Note that each tutorial decision is clear, explicit, and quite separate from any book-keeping information associated with teaching performance on which tutorial decisions are not dependent.

Expressing task analyses

The example given above is essentially a simple tutorial strategy, with a task analysis being implicit in the tutorial situations in which the various teaching operations are applied. The partitions on LASTOP, STATE and SCORE illustrate how classes of tutorial situation are distinguished. The partitions on NUMOP are examples of thresholds.

Where CAI programs have a non-trivial task analysis, such analyses typically take one of two forms:- a task difficulty model or an ordering of concepts or teaching subgoals.

Task difficulty models can be expressed as a system of production rules where the elements of the state-vector are dependent on possible student states of knowledge and measures of difficulty on the teaching material. In this case the actions correspond to potential features of components of the teaching material. Such a system would be closely analogous to the set of productions involving classes and thresholds illustrated in the above example.

On the other hand, an ordering of teaching subgoals can be expressed as a branching network of the type used by Smallwood (1962). Such networks can be translated into a system of production rules by introducing an element (the goal-element) of the state-vector in which the current tutorial subgoal is stored. Partitions can then be defined to detect the currently active teaching subgoals. This scheme does, however, necessitate the introduction of a class of dummy teaching actions. These dummy actions set up the currently active teaching subgoals. The execution of such an action will not directly result in any teaching occurring and its occurrence in a condition-action rule can be regarded as a 'comment' or 'annotation' on the likely intention involved in executing the other actions on the right-hand side of the rule. The use of such devices to impose a higher level of control on the execution of a set of production rules has been criticised by Davis and King (1975) as "not in the spirit" of production rule programming. However, such devices are of great utility if there are different levels of organisation inherent in the task being executed by a set of productions, and similar devices have been strongly advocated by various workers who have used rule-based systems for scientific discovery (see, for example, Lenat and Harris, 1977). This deviation from the standard simple form of control can be mitigated by clearly marking those actions which affect the flow of control. The values of the goal-element will correspond to the node set of the network. Consider the network given in figure 4. Taking the state-vector with elements NODE and CONDITION, and where every partition has the form $X \rightarrow XI$ if $X = XI$, then the network translates into the following set of productions:

(START, C1) → B1
(START, C2) → B2
(B1, C3) → B3
(B2, C5) → B4
(B2, C4) → B3
(B4, C4) → START
(*, C6) → STOP

The actions correspond to presenting whatever teaching material is associated with the note. Each rule except the last corresponds

Figure 4: A branching network

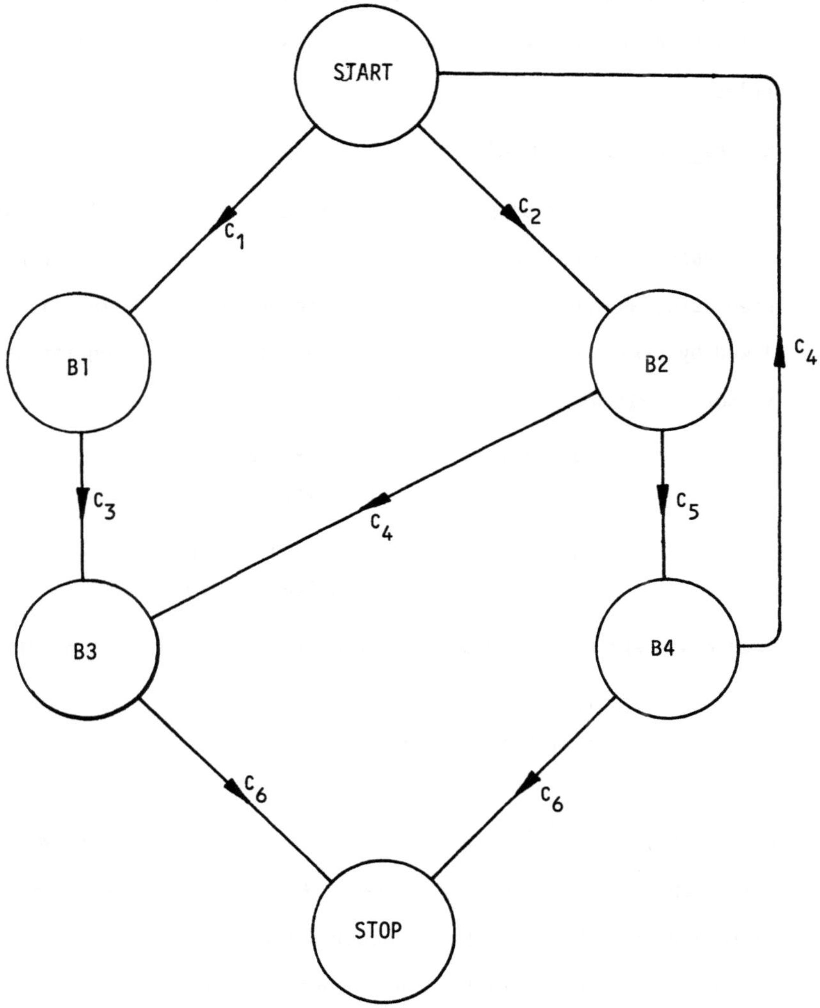

Note: C_i correspond to conditions which must be fulfilled for arc to be followed.

to an individual arc. The last rule corresponds to two arcs. In an ordered set of rules it is possible to detect redundant rules (that is, rules which are completely subsumed by other rules). Note, however, that in creating the last rule some information has been lost, namely the antecedent nodes to the STOP node. This could be a source of difficulty in the course of automatic evaluation or manipulation of these rules.

Expressing student models

Consider two simple student models to predict whether a student has grasped the concept 'prime number'. First suppose a student has been asked whether he knows what a prime number is and this is followed by asking him whether a number 'N' is prime. Using the state-vector (FAMILIAR, ANSWER) and the partitions

$$\text{FAMILIAR} \rightarrow \text{A1 if FAMILIAR = YES}$$
$$\rightarrow \text{A2 if FAMILIAR = NO}$$

$$\text{ANSWER} \rightarrow \text{B1 if ANSWER = CORRECT}$$
$$\rightarrow \text{B2 if ANSWER = WRONG}$$

then a student model could be given by the condition-action rules:

$$(\text{A1}, \text{B1}) \rightarrow 1$$
$$(\text{A2}, \text{B1}) \rightarrow f(\text{N})$$
$$(*, \text{B2}) \rightarrow 0$$

where $f(N) = 1 - \frac{P(N)}{N}$ if N is prime and $f(N) = \frac{P(N)}{N}$ if N is not prime, and $P(N)$ = Number of primes $\leq N$. The action is a measure of certainty that the student has acquired the concept. Such a measure of certainty could be used as an element of a state-vector and could be employed to help determine an appropriate teaching operation or elicit further information about the student's state of knowledge (this would involve changing the definition of $f(N)$ above).

An alternative approach would be to simulate the various ways in which the student is performing a task. For example, suppose in the prime number example it is hypothesised that students either (a) guess, (b) believe that all odd numbers are prime, or (c) determine whether a number N is prime by successively dividing it by all numbers M where $1 < M < N$. Taking as state-vector (LASTDIVISOR, REMAINDER) and partitions:

```
LASTDIVISOR  →  A1   LASTDIVISOR = 2
             →  A2   LASTDIVISOR = N - 1

REMAINDER    →  B1   REMAINDER ≠ 0
             →  B2   REMAINDER = 0
```

then three alternative sets of condition-action rules to model the three classes of student could be as follows (with LASTDIVISOR = 1 initially):

```
a)  (*, *)    →  RANDOM GUESS

b)  (A1, B1)  →  SAY 'PRIME'
    (A1, B2)  →  SAY 'NOT PRIME'
    (*, *)    →  DIVIDE N BY 2

c)  (A2, B1)  →  SAY 'PRIME'
    (A2, B2)  →  SAY 'NOT PRIME'
    (*, *)    →  DIVIDE N BY LASTDIVISOR + 1
```

Such a set of alternative student models could be used in conjunction with a tutorial strategy by 'running' each model on the material presented to the student and comparing the performance of each student model with actual student performance.

Expressing complex tutorial strategies with production rules

We may classify* the ways of organising systems of production rules as follows:

a) Single or multiple action - whether the right-hand side of the production rule is restricted to a single action or not.

b) Single or multiple hit - whether a single condition evaluating true is sufficient for a rule to fire, or whether all the conditions have to evaluate true.

c) Ordered or unordered - whether the mechanism employed to resolve conflicts when one or more rules fires is ordering of the rules.

d) Single or multiple cycles - whether each right-hand side executed results in some change in the external environment as opposed to all the actions on a right-hand side resulting in internal changes (for example, setting flags or declaring goals).

e) Flat or nested - whether after execution of a right-hand side control always returns to the set of productions currently being executed as opposed to allowing actions themselves to either constitute or invoke other sets of productions.

The simple example expressing a tutorial strategy given above was single action, multiple hit, ordered, single cycle and flat. To express an entire complex tutorial strategy in this form would result in a very large state-vector, lots of parsing, and large numbers of cumbersome condition-action rules with much redundancy. This results in inefficiency, both in terms of processing time and storage space. The following design choices can be employed to gain efficiency and simpler, more transparent structures:

---

* Based on a private communication of D.A. Waterman, 1973.

a) The use of multiple actions where executing one action automatically entails the execution of another.

b) The use of single hits as a way of expressing a logical 'or'. In a context with few actions and a large number of sparse condition-action rules, a marker can be used to denote 'or' (i.e., any single hit constitutes a match). This reduces the number of rules required.

c) Careful choice of order of rules can reduce the number of condition-action rules needed (cf. branching network example).

d) Single cycle operation can reduce the amount of processing required and is possible if multiple actions are employed because 'internal actions' (e.g. declaring goals) can then be associated with (i.e. used on the same right-hand side as) the appropriate external actions.

e) Nesting can be used to implement hierarchical strategies. Consider a set of condition-action rules representing a set of means-end guidance rules. The actions could result in the execution of a set of production expressing a student model or task analysis; that is, production rules can be used as a switch between subsets of the entire set of productions. Each subset may be regarded as a production rule 'subroutine'.

All the above techniques are applied in the specific implementation discussed in the following chapter.

3.2 Changing tutorial strategies

The principal rationale for representing tutorial strategies as sets of production rules is that such a representation is amenable to automatic manipulation. Now the number of possible changes which may be executed on a non-trivial set of production rules is very large. However, the changes which are meaningful within the context of experimenting with a tutorial strategy are fewer and we will only consider those which can be interpreted as:

a) Changing the action employed in a class of tutorial situations.

b) Changing the definition of a class of tutorial situations.

c) Creating a new definition for a class of tutorial situations.

d) Altering thresholds for the execution or inhibition of some teaching operation.

e) Changing the order in which tutorial subgoals are adopted.

Note that in terms of actual teaching performance the above five types of change are not necessarily distinct. Both (d) and (e) are special instances of (b). The execution of (b) or (c) will, if a non-trivial change has been made, result in one or more occurrences of (a).

The above types of change will be illustrated with a simple example. Consider the fourth condition-action rule given in the earlier example set of production rules on page 42

$$(\text{LASTOP1}, *, \text{SCORE1}, \text{NUMOP1}) \rightarrow T(2) .$$

This rule is matched and T(2) executed if the symbolic state-vector does not match any of the three rules above this rule, and if the symbols LASTOP1, SCORE1, and NUMOP1 are present in the appropriate positions in the state-vector; i.e., if LASTOP = T(1), SCORE $\leqslant$ 5 and NUMOP > 4. For the purposes of this discussion, symbols defined by partitions such as NUMOP > 4 will be referred to as 'restrictions' on the values of the relevant elements of the state-vector. With respect to the rule given above the set of production rules can be changed as follows:

a) By changing the action on the right-hand side, for example

(LASTOP1, *, SCORE1; NUMOP1) → T(1)

b) By changing the restrictions on the left-hand side, for example

(LASTOP1, *, SCORE2, NUMOP1) → T(2)

Alternatively a restriction could be added to or dropped from the left-hand side.

c) By defining new restrictions, for example

SCORE → SCORE3 if SCORE $\leqslant$ 3
NUMOP → NUMOP4 if NUMOP > 2

and inserting the rule

(LASTOP1, *, SCORE3, NUMOP4) → T(2)

above

(LASTOP2, *, SCORE1, NUMOP1) → T(2) .

d) Altering a threshold, for example by changing

NUMOP → NUMOP1 if NUMOP > 4

to

NUMOP → NUMOP1 if NUMOP > 6

e) The order in which tutorial subgoals are adopted could be changed by rewriting the fourth rule either as in (a)

(LASTOP1, *, SCORE1, NUMOP1) → T(1)

or by changing a restriction concerned with previous tutorial subgoals, for example, by rewriting the fourth rule as follows:

(LASTOP2, *, SCORE1, NUMOP1) → T(2) .

A production rule amender

In order to make the above ways of changing production rules more precise, we present a design for a production rule amender (a substantial part of this amender has been implemented and is discussed further in chapter 4).

The operation of the amender involves the execution of two types of function: (a) predicates which identify the production rules to be changed, and (b) editing functions which execute the changes. The functions are:

(a)(i) PRESENT, ABSENT (PRESENT *x*)* identifies a condition-action rule, if *x* identifies an element of the state-vector on which a restriction is placed or if *x* is a member of the list of actions on the right-hand side; for example (LASTOP1, *, SCORE1, NUMOP1) → T(1), (PRESENT ≡LASTOP) returns ≡true. ABSENT is similarly defined.

(a)(ii) LESS, GREATER, EQUAL, NOTEQUAL (LESS *x* *y*) identifies a condition-action rule if (PRESENT *x*) returns true and if the restriction on *x* is such that it is less than *y* numerically. For example, for (LASTOP1, *, SCORE1, NUMOP1) → T(1) (LESS ≡SCORE 6) is true because SCORE1 is defined as SCORE → SCORE1 if SCORE ≤ 5. GREATER, EQUAL and NOTEQUAL are similarly defined.

(a)(iii) INCLUDES, EXCLUDES (INCLUDES *x* *y*) identifies condition-action rules if (PRESENT *x*) is true and if the restriction on *x* is such that it must be a member of the set *y*. For example, if we rewrite STATE → STATE1 if STATE = C(1) or STATE = C(2) as the equivalent STATE → STATE1 if (MEMBER STATE ≡(C(1) C(2))). Then for (*, STATE1,

---

* LISP notation (C. Weissman, 1967) is used throughout. (PRESENT *x*) is interpreted as apply the function PRESENT to the value of the parameter *x*. The prefix ≡ indicates a literal atom.

*, *) → T(3) (INCLUDES ≡STATE ≡(C(1) C(3))) returns false and (INCLUDES ≡STATE ≡(C(1) C(2) C(3))) returns true. EXCLUDES is similarly defined.

(a)(iv) BEFORE, AFTER (BEFORE *x*) may be interpreted in two ways. If *x* is a tutorial subgoal, (BEFORE *x*) identifies the condition-action rules which correspond to the incoming arcs of the node corresponding to the subgoal. For example, for the rule

(LASTOP1, *, SCORE1, NUMOP1) → T(2)

(BEFORE ≡T(2)) and (AFTER ≡T(1)) are true. The latter is true as a result of the restriction on LASTOP given by LASTOP1, namely LASTOP → LASTOP1 if LASTOP = T(1).

If *x* is an action, (BEFORE *x*) identifies the position immediately before *x* in the sequence of actions on the right-hand side of the condition-action rule. AFTER is similarly defined.

(b)(i) RAISE, LOWER (RAISE *x*) is interpreted as raise the threshold or parameter identified by *x*. The magnitude of the increase could be determined by a parameter associated with the running of experiments. Alternatively, the magnitude of the change could be indicated by a second parameter, e.g. (RAISE ≡SCORE 2) where the condition-action rule to be changed has been identified as:

(LASTOP1, *, SCORE1, NUMOP1) → T(2).

The result of executing the change is the creation of the partition

SCORE → SCORE3 if SCORE ≤ 7

and rewriting the rule as

(LASTOP1, *, SCORE3, NUMOP1) → T(2).

Note that the change cannot be executed by rewriting

SCORE → SCORE1 if SCORE ≤ 7,

as the restriction SCORE1 is also employed in another rule. LOWER is similarly defined.

<u>(b)(ii) WIDEN, NARROW</u> These functions are similar to RAISE and LOWER, but apply to the situation where a restriction is defined as a partition on a range of numerical values by employing two boundary conditions, for example,

NUMOP → NUMOP2 if 10 > NUMOP > 7 .

In this circumstance a parameter must be used to identify which boundary condition is to be altered. For example, (WIDEN ≡NUMOP ≡LOWER 2) if executed on the above partition results in

NUMOP → NUMOP2 if 10 > NUMOP > 5 .

NARROW is similarly defined.

<u>(b)(iii) ADD, DELETE</u> (ADD x y) is executed by changing the subset restriction on x by adding the element y to that subset. For example, executing (ADD ≡STATE ≡C(3))

STATE → STATE1 if (MEMBER STATE ≡(C(1) C(2)))

results in the rewritten partition

STATE → STATE1 if (MEMBER STATE ≡(C(1) C(2) C(3))) .

DELETE is similarly defined.

<u>(b)(iv) INSERT, REMOVE, REPLACE</u> These functions are defined with respect to the conditions or actions component of a set of condition-action rules and also with respect to tutorial subgoals. In the first case (INSERT x) is interpreted as adding the appropriate restriction, as defined by x , to the left-hand side, or adding the action x to the right-hand side. For example, executing (INSERT (GREATER NUMOP 6)) on

(*, STATE1, *, *) → T(3)

results in creating the partition

NUMOP → NUMOP4 if NUMOP > 6

and rewriting the condition-action rule as:

(*, STATE1, *, NUMOP4) → T(3) .

In the case where the ordering of tutorial subgoals is being changed, the operation of the three functions is illustrated in figure 5. For example, executing (AFTER ≡START)(BEFORE ≡B1)(INSERT ≡B5) on the set of condition-action rules derived from figure 4 (see page 45), results in rewriting

(START, C1) → B1 as (START, C2) → B5

and creating the new rule

(B5, C1) → B1 .

(REMOVE x) and (REPLACE x) are defined similarly, except that their invocation results in the automatic execution of both (BEFORE x) and (AFTER x) to identify the condition-action rules to be changed. Execution of (REMOVE x) should also have the effect of deleting all the actions following x associated with that particular tutorial subgoal in the rule containing x .

Operation of amender

The amender operates on a list of calls to the above functions. First, predicates are executed to identify the condition-action rules which are candidates for amendment. Of these candidates, those which are changeable by the editing functions can be determined as follows. Where the arguments of these functions are not concerned with tutorial subgoals they can be interpreted as PRESENT, except for ADD and INSERT which can be interpreted as ABSENT. Where these arguments are concerned with tutorial subgoals then REMOVE and REPLACE can be interpreted as noted in (b)(iv).

<u>Figure 5</u>: Example of operation of amender on groups of production rules expressing orderings of tutorial subgoals.

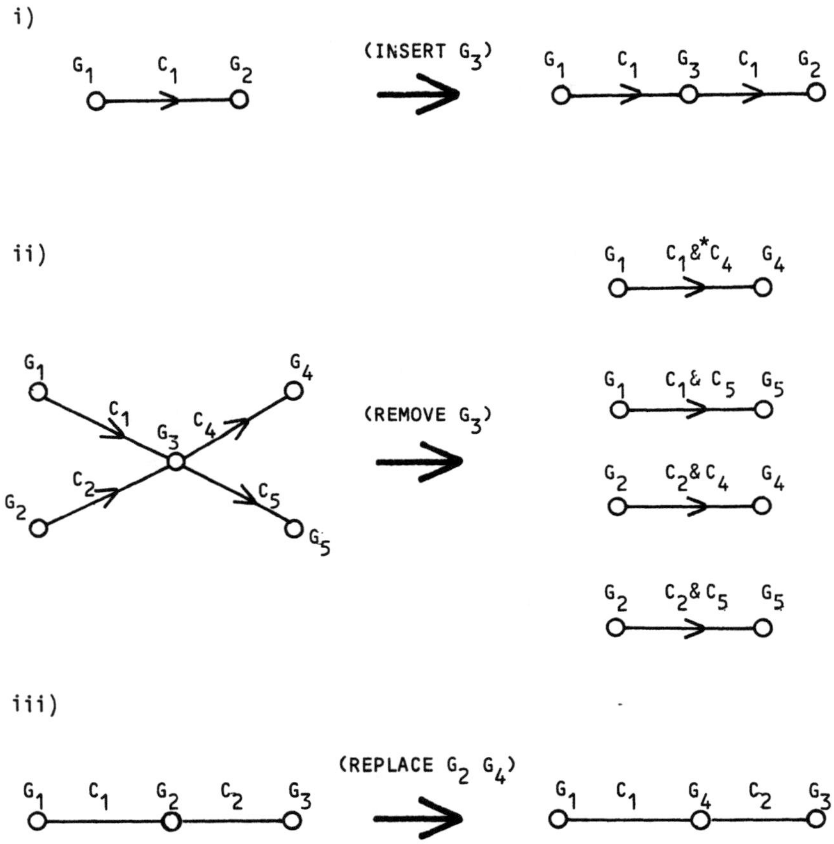

* $C_I \& C_J$ stands for the condition (C_i AND C_j) with the restriction that if $C_i \& C_j$ is impossible, C_i takes precedence over C_j.

<u>Note</u>: G_i are subgoals. C_i are conditions on the subgoals and correspond to the condition part of condition-action rules.

Having thus eliminated rules which are not potentially changeable, the editing functions can be executed, thus changing the set of production rules. When changing the condition part of a condition-action rule the following procedure is adopted. First, if an appropriate restriction has already been defined in the set of partitions, then employ it. If this is not the case, but the restriction to be changed only occurs in the condition-action rule to be changed and other rules to be changed in the same way, then change the definition of the restriction appropriately. Otherwise, if neither of the above is possible, create a new definition of an appropriate restriction and employ this. This simple mode of operation is feasible because the restrictions employed are either partitions on the integers or subset relations.

Side effects

Changing a condition-action rule to achieve a particular goal may very well result in 'side effects'. That is, unpredicted and possible unwanted changes in the application of other condition-action rules may occur. There are two sorts of ways side-effects can originate: (a) As a result of the way the set of production rules is organised. For example, a symbolic state-vector previously reaching and matching a given condition-action rule (in an ordered set of rules) may no longer reach this rule as a result of a change in the condition part of a rule above it. (b) As a result of the incompletely known properties of some teaching operation. For example, if there is a change in the action executed in a given context (or change in the context) the state-vector may never have a set of values, which sometimes occurred prior to the change. Again the result is that a rule previously applied may no longer be applied or a rule previously not applied may now be applied.

Now the occurrence of side effects will not necessarily be a problem. After the execution of a change, the whole set of production rules will be evaluated by statistical inference on measures of subsequent performance. This makes it possible to undo changes with 'bad' side effects. In fact, as the system is running experiments on hypotheses derived from an incomplete theory of instruction (if the theory were complete it should not be necessary to run experiments and ought to be possible to construct an 'optimal' teaching program with respect to that theory - see Atkinson, 1972), it would be very surprising if there were no unpredicted side effects. The occurrence of unwanted side effects of type (a) can be reduced by carefully organising sets of production rules and by the use of restrictions which, although redundant, are known to be contingent on a condition-action rule being matched. For example, in the set of production rules on page 42 in the seventh rule

$$(*, *, SCORE2, *) \rightarrow T(2)$$

the symbol SCORE2 is strictly speaking redundant as SCORE2 is defined by:

$$SCORE \rightarrow SCORE\ 2 \quad \text{if} \quad SCORE > 5$$

and the rule immediately above it is:

$$(*, *, SCORE1, *) \rightarrow T(1)$$

where SCORE1 is defined by:

$$SCORE \rightarrow SCORE1 \quad \text{if} \quad SCORE \leqslant 5 .$$

But writing the seventh rule as

$$(*, *, *, *) \rightarrow T(2)$$

could result in problems if the definitions of SCORE1 were subsequently changed.

The extent to which side effects of type (b) will occur depends on the theory of instruction. As the program runs experiments and detects these side effects, the circumstances under which they occur should become incorporated into the theory of instructions and thus no longer be unpredicted. (This presupposes the ability to learn and generalise from the results of experiments. This is discussed further in 3.3.)

Now, while the majority of changes and their associated side effects will have to be evaluated experimentally, it is possible to detect without running experiments some changes which, when viewed in the context of the entire set of production rules, will certainly result in a severe deterioration of teaching performance. A prerequisite for this is a definition of 'acceptable' teaching performance in terms of explicit constraints on teaching performance.

Note, however, that side effects are bound to occur. It may be possible to filter out some of the clearly disastrous changes and monitoring with respect to measures of acceptable performance may be possible during teaching. However, the principal and usual way of evaluating tutorial strategies expressed in sets of production rules will be experiment and subsequent statistical inference. This experimental methodology is the basis of the design given in chapter 2, and provides the principal safeguard against repeatedly changing the strategy in such a way that a permanent deterioration in teaching performance results.

<u>Redundancy</u>

Another sort of side effect that occurs is that rules may become redundant. Accordingly, for efficiency, redundant rules could be identified and removed at intervals. The following may be applicable:

a) If a restriction is not employed in the condition part of any condition-action rule, delete the partition defining it.

b) Condition-action rules which are not applied during very long periods of teaching performance could be deleted.

c) It may be possible to identify condition-action rules which will never be applied as a result of the ordering of the system of production rules.

<u>Assumptions implicit in design of amender</u>

There are a number of assumptions implicit in the design of this amender. Most of these stem from the general assumption that the set of production rules being amended expresses a tutorial strategy (or component of one). The assumptions are:

a) An index or directory is provided which identifies the following classes of objects on which the amender operates:

   (i) The elements of the state-vector.
   (ii) The elements of the state-vector related to tutorial subgoals.
   (iii) The possible actions.
   (iv) The actions associated with particular tutorial subgoals.

b) The values of the elements of the state-vector will either be numbers or a list of names.

c) Partitions on numerical elements will be expressed with the relations $>$, $\geq$, $=$, $<$, $\leq$ and numerical constants. So for example

$$\text{NUM} \rightarrow \text{A1} \quad \text{if} \quad \text{NUM} > \text{NUM}^2 - 7$$

or

$$\text{NUM} \rightarrow \text{A2} \quad \text{if} \quad \text{NUM even}$$

are not permissible partitions.

d) Partitions on elements involving lists will be expressed in terms of membership or non-membership of some subset of the total possible set of names, where such subsets are defined by enumerating their elements.

Now, assumption (a) is not restrictive and makes it easier to implement an amender. For example, (a)(ii) makes it possible to distinguish which of the two interpretations of (AFTER x) is intended (that is, is a list of actions or a directed graph of subgoals being operated on).

Assumptions (b), (c), and (d) ensure that partitions have a comparatively straightforward 'logic' and can be compared to see if they are equivalent, overlap, disjoint, or if one partition subsumes another. This of course facilitates amending and non-experimental evaluation. At the same time these three assumptions pose restrictions on the form in which tutorial strategies can be expressed. Thus (b) excludes any data type for the value of the element of a state-vector except numbers or lists of names. There are situations in which it might be desirable for the value of an element to be, say, an array or a directed graph. For example, it may be appropriate to represent a student's progress by a subgraph representing that portion of the teaching material the student has been exposed to. Similarly, it might be desirable to define a partition on a numerical element by a function, say, testing whether the element is prime. Or it could be useful to define a partition using the order of the elements of a list. The use of such functions is precluded by (c) and (d).

In the course of this study the restrictions presented by (b), (c), and (d) have posed no problems. If they were to pose a serious problem, they might be circumvented as follows:

Firstly, augment the logic of the amender to handle the new data type or new function to be used in the definition of partitions. Then introduce a new element to the state-vector, and introduce a new teaching operation which operates on the new data-type or incorporates the function to be used to define the partition and update the new element of the state-vector as a result of executing this operation.

For instance, suppose it is desirable to test whether a number x is prime. Then introduce an element of the state-vector called xPRIME. Define a function PRIMETEST which tests whether x is prime and updates xPRIME accordingly (i.e., xPRIME = "PRIME" or xPRIME = "NOTPRIME"). Then after any teaching operation involving the generation of a new x insert a call to PRIMETEST.

3.3 Representing theories of instruction

The previous section described an amender for tutorial strategies expressed in terms of production rules. This section is concerned with the problem of identifying potentially useful changes (that is, those which result in some goal variable related to teaching performance being maximised) for the amender to execute. One solution to this problem would be to employ a list of specific calls to the amender. This approach would be limited. The results of trying a specific call could not be applied to evaluating the other calls or to generating a new potentially useful call. Further, such a list would be excessively long.

The approach adopted here is based on the assumption that explicit goals defining improvement in teaching performance can be identified for a given teaching situation. Then, from the theory of instruction for the particular teaching context, assertions are determined by the implementer relating goals to aspects of tutorial strategy. These assertions form a data base on which a deduction procedure operates deducing hypotheses for potentially useful changes. This section is concerned with the development of a deduction system incorporating an appropriate form for such assertions, and the following section describes a suitable deduction procedure to operate on these assertions. To fulfil the above requirements and those specified by the design (see 2.5) the deduction system should have the following characteristics:

a) It should be possible to write assertions expressing information relevant to the goals.

b) It should be possible to automatically identify changes likely to result in the achievement of goals using these assertions.

c) It should be possible to automatically rewrite or create new assertions as a result of carrying out an experiment by executing a change on the tutorial strategy. Ideally these new expressions should constitute reasonable generalisations of the results of the experiment.

d) It should be feasible to give 'advice' resulting in the modification or creation of new expressions. This advice need not necessarily take the form of English statement. However, no substantial re-programming should be necessary to get the system to accept advice. In practice it is envisaged that advice could take the form of additional assertions.

e) The deduction procedure should be capable of operating in conjunction with a CAI program running in real-time.

The deduction system adopted in this study is *ad hoc*, developed in response to the particular problem area under consideration, and limited in scope by 'historical' reasons such as time and the software facilities available.

It is a development of Black's (1968) natural deduction system. Two types of assertion are employed. Firstly causal assertions which relate actions (changes executable on the set of production rules embodying the teaching strategy) to effects (possible changes in teaching performance). Secondly, definitional assertions which group together classes of actions or effects. The set of causal assertions can be viewed as an action-driven production system (see Waterman, 1977), and is analogous to a collection of PLANNER consequent theorems (see Hewitt, 1972). The set of definitional assertions is essentially a list of associations used during the matching process to prevent endless deductions and facilitate generalisations (see 3.4).

<u>Causal assertions</u>

These are of the general form (OPERATOR ACTIONS EFFECTS). The OPERATOR qualifying the assertion may be either 'CERT' or 'POSS'. ACTIONS is of the form (Connective $A_1 \ldots A_k$). A_i must be of the form (ACTION $N_1 \ldots N_\ell$). ACTION will be regarded as a predicate and executable (that is, a call to the amender). N_i must be a name (that is, not a predicate applied to a name). CONNECTIVE is a logical 'or' or a logical 'and' and is omitted if $\ell = 1$. EFFECTS is of the form (CONNECTIVE $B_1 \ldots B_m$) where B_i must be of the form (EFFECT $N_1 \ldots N_n$) and EFFECT is a predicate but not executable and N_i are as above. The CERT and POSS operators are employed to distinguish consequences which will certainly follow actions from consequences which may follow actions.

Examples of causal assertions for a hypothetical teaching context where the goals are to maximise a variable SCORE and minimise a variable TIME are:-

1. (CERT (SHORTEN SESSIONS) (DECREASE TIME))

2. (POSS (SHORTEN SESSIONS) (DECREASE SCORE))

3. (POSS (REPLACE OPERATION1 OPERATION2)(OR(DECREASE TIME) (INCREASE SCORE)))

4. (CERT (AND(LOWER THRESHOLD1)(LOWER THRESHOLD2))(SHORTEN SESSIONS))

These assertions may be interpreted as follows:-

1. "For any set of production rules expressing a tutorial strategy for this context shortening sessions will result in a lowering in value of the variable TIME."

2. "For some sets of production rules expressing a tutorial strategy for this context, shortening sessions will result in a decrease in the value of the variable SCORE."

3 and 4 are interpreted similarly. Note that three classes of predicate can be distinguished.

i) Those whose arguments are 'goal variables' such as (DECREASE TIME) or (INCREASE SCORE).

ii) Those which are potentially executable such as (REPLACE OPERATION1 OPERATION2) or (LOWER THRESHOLD1).

iii) Those which are constructs and make it possible to identify classes of action or effect and introduce useful concepts such as (SHORTEN SESSIONS). Note that if shortening sessions had n different effects on goal variables and there were m different actions which resulted in shortening sessions, then this could be expressed in $m+n$ assertions with the construct (SHORTEN SESSIONS) but it would require $m \times n$ assertions without this construct.

Deductions can be made from such assertions (see 3.4) rather in the fashion employed by Black's (1968) deductive question-answering system. The assertions can be used to determine which actions should be employed to achieve certain desired effects. Alternatively, they can be used to determine all the possible effects of certain actions. Consider the four example assertions given above and some of the deductions which can be made from them. For example, with the goal of (DECREASE TIME) then by applying assertion 1 it can be deduced that an action certain to cause this is (SHORTEN SESSIONS). Then from 4 it is certain that (AND(LOWER THRESHOLD1)(LOWER THRESHOLD2)) will result in (SHORTEN SESSIONS). As (LOWER x) is executable, a strategy which will certainly result in (DECREASE TIME) has been identified. But note that assertion 2 can be applied to deduce that, as a result of (SHORTEN SESSIONS), the effect of executing the left-hand side of 4, there is a possibility that (DECREASE SCORE), an undesirable side-effect, will occur. However, by applying assertion 2, it can be deduced that executing (REPLACE OPERATION1 OPERATION2) may possibly result in the goal (DECREASE TIME) and in addition may possibly have the desirable side-effect (INCREASE SCORE).

Thus in this simple example two alternative courses of action are identified, one certain to achieve the goal but which can however have a 'bad' side-effect, and one which may possibly achieve the goal and which may have a 'good' side-effect. The types of heuristics which can be employed to discriminate between possible courses of action in such situations are discussed below with the detailed description of the deduction procedure.

<u>Definitional assertions</u>

These define names (as opposed to causal assertions which relate actions to effects). They are employed in the deduction procedure to test for the equivalence of clauses in causal assertions. They are of the form (DOPERATOR CLASS CLASS), where DOPERATOR is either 'DEFCERT', 'DEFPOSS' or 'DEFEQUIV'. CLASS has the form (CONNECTIVE $M_1 \ldots M_p$) and M_i is the name of a variable or of a predicate. Definitional assertions are used to define classes or possible classes of variables, actions or effects which are equivalent (or possibly equivalent) in some way. Examples of definitional assertions are:

1. (DEFCERT (OR INCREASE DECREASE) CHANGE)
2. (DEFPOSS CHANGE (OR INCREASE DECREASE))
3. (DEFEQUIV (AND THRESHOLD1 THRESHOLD2) THRESHOLD).

1. This assertion may be interpreted as "an INCREASE OR DECREASE is a CHANGE". Given this assertion, then in any clause in the EFFECTS part of a causal assertion, INCREASE or DECREASE can be rewritten as CHANGE without affecting the validity of that causal assertion. For example, from (CERT ACTION1 (INCREASE *x*)) and the definitional assertion it is possible to deduce the causal assertion (CERT ACTION1 (CHANGE *x*)). Similarly, in any clause in the ACTIONS part, CHANGE can be rewritten as INCREASE or DECREASE. For example, from (1) and (POSS (CHANGE *x*) EFFECT1) it is possible to deduce (POSS (INCREASE *x*) EFFECT1) and (POSS (DECREASE *x*) EFFECT1). Such deductions are not explicitly carried out but are implicit in the matching algorithm employed by the deduction procedure (see 3.4).

2. This assertion may be interpreted as "a CHANGE is either an INCREASE or a DECREASE". Given this assertion, then in any clause in the EFFECTS part of a causal assertion, CHANGE can be rewritten as INCREASE or DECREASE, possibly affecting the validity of that assertion. For example, from (CERT ACTION1 (CHANGE *x*)) and definitional assertion (2) it is possible to deduce (POSS ACTION1 (INCREASE *x*)). Similarly in the ACTIONS part of a causal assertion, INCREASE or

DECREASE can be rewritten as CHANGE, possibly affecting the validity of that assertion. For example, from (POSS (INCREASE x) EFFECT1) and definitional assertion (2) it is possible to deduce (POSS2 (CHANGE x) EFFECT1), where POSS2 stands for 'possibly possibly'. (Various interpretations of 'possibly possibly' are given in 3.4.)

3. This assertion may be interpreted as "THRESHOLD is identical to THRESHOLD1 and THRESHOLD2 and *vice versa*". That is, (DEFEQUIV x y) is equivalent to (DEFCERT x y) and (DEFCERT y x). Thus any clause with THRESHOLD can be rewritten with THRESHOLD1 and THRESHOLD2 and *vice versa*. For example, (LOWER THRESHOLD) is equivalent to (AND (LOWER THRESHOLD1) (LOWER THRESHOLD2)).

The use of definitional assertions reduces the number of causal assertions needed. Where a causal assertion is true in a number of instances a name can be defined. The use of definitional assertions also removes the possibility of endless deductions, in the deduction procedure, as a result of repeated substitution. (For a discussion of endless deduction in such systems, see Black, 1968.) Definitional assertions are used in the deduction procedure to match the arguments of causal assertions, but are not substituted for them. Suppose the following were causal assertions:

1. (POSS (CHANGE x)(INCREASE x)
2. (CERT (INCREASE x) (CHANGE x)).

With the goal (INCREASE x) from (1) we deduce that (CHANGE x) may possibly achieve this goal, then from (a) we deduce that to (CHANGE x) we certainly will succeed if we achieve (INCREASE x), then from (1) etc., etc. However, in the formulation given, such assertions are of type 'definitional'. Provided that definitional assertions are created where appropriate by the user, it will not be possible for repeated substitutions to occur. But the deduction procedure is not robust, and a single well-formed but nonsensical causal assertion (for example, one with an executable predicate on the right-hand side) could cause endless deductions to occur.

Examples of assertions

The principal example of a set of assertions given in this thesis is the set of assertions used in the experimental program and listed in Appendix VI. The examples given below have been devised to represent possible components of a theory of instruction with respect to some of the CAI programs discussed in chapters 1 and 2. The assumption is made that in each case the tutorial strategy has been expressed as a set of production rules. In the examples below the prefix SV indicates that the variable concerned is an element of the appropriate state-vector. The prefix * indicates that a predicate is potentially executable.

The first example is for Kimball's integration program (see 1.2), assuming goals of minimising a variable TIME and maximising a variable SCORE.

1. (DEFCERT THRESHOLD (OR PROBLEM PROBLEMLENGTH TECHNIQUE))
2. (POSS (*LOWER THRESHOLD) (DECREASE TIME))
3. (CERT (*LOWER THRESHOLD) (INCREASE INTERRUPTIONS))
4. (POSS (AND(INCREASE INTERRUPTIONS) (FAST STUDENT))
   (DECREASE SCORE))
5. (POSS (AND(INCREASE INTERRUPTIONS) (SLOW STUDENT))
   (INCREASE SCORE))
6. (CERT (*GREATER SVSCORE 30) (FAST STUDENT))
7. (CERT (*LESS SVSCORE 10) (SLOW STUDENT)
8. (CERT (AND(*GREATER SVSCORE 10) (*LESS SVSCORE 30))
   (OR(FAST STUDENT) (SLOW STUDENT))) .

Further assertions could be written qualifying the potential effects of changing the different thresholds defined by (1). In the absence of (6), (7) and (8) it would be possible to read assertions to change the definition of 'fast' or 'slow' student.

The second example is for the Leeds Arithmetic (see 1.2), assuming goals of maximising RL, rate of learning, and maximising M, motivation. SVPS stands for the pupil's probability of success. It is assumed that measures of the pupil's industry and intelligence are available.

1. (DEFEQUIV SCORE (AND RL M))
2. (POSS (BORE STUDENT) (DECREASE SCORE))
3. (POSS (*RAISE SVPS) (BORE STUDENT))
4. (POSS (AND (*RAISE SVPS) (STUDENT LAZY)) (DECREASE RL))
5. (POSS (AND (*RAISE SVPS) (STUDENT INDUSTRIOUS)) (INCREASE RL))
6. (POSS (AND (*RAISE SVPS) (STUDENT INTELLIGENT)) (BORE STUDENT))
7. (POSS (AND (*RAISE SVPS) (STUDENT UNINTELLIGENT)) (INCREASE M))

Again further assertions could be added. From the above examples it is readily apparent that as a form for expressing theories of instruction the assertions used are limited. In particular, it is not possible to express mathematical models of learning except in a very gross 'action x causes variable y to increase' fashion. However, such assertions can be used to identify both particular and general possibly useful changes in tutorial strategy.

3.4 A deduction procedure for generating hypotheses

There are two sorts of deductions which can be usefully made on a set of assertions expressing a theory of instruction.

Firstly, given a desired goal such as 'decrease computer time', deduce the sets of actions (changes in tutorial strategy) which may achieve this goal.

Secondly, given a set of actions, deduce the possible consequences of these actions.

If alternative sets of actions are to be compared, then some measure of the likelihood of a consequence occurring is necessary. Here is a deduction procedure (operating on assertions in the form described in 3.3) which computes possible consequences of different courses of action:-

1. Select a goal, for example (DECREASE (COMPUTER TIME)).
2. Perform a search for matches on the right-hand sides of the causal assertions with this goal. The definitional assertions are used to check for matches, e.g. with the assertion (DEFEQUIV TIME (OR (STUDENT TIME) (COMPUTER TIME))) the goal of (DECREASE (COMPUTER TIME)) will match (CERT ACTION (DECREASE TIME)).
3. The corresponding left-hand sides are detached, and Step 2 is repeated with any non-executable predicates as goals. Thus a list of executable predicates corresponding to function calls to the amender is built up.
4. For the lists $<L_i>$ of executable predicates thus obtained, POSS(L_i), the number of causal assertions qualified by the POSS operator employed in the deduction of $<L_i>$ is computed. Any list containing a non-executable predicate which cannot be matched on the right-hand side of some causal assertion is discarded. The α $(\alpha \geq 2)$ lists $<L_i>$ with the smallest value of POSS(L_i) (i.e. the most likely) are taken as possible courses of action.
5. For each list of executable predicates, $<L_i>$ from $<L_i>_\alpha$ matches are made on the left-hand side of causal assertions. The corresponding right-hand sides are

detached and again matched with left-hand sides. This process is repeated until no new matches can be made.

6. From (5) all the predicates $<M_j>$ where arguments are 'goal' variables are identified. These are divided into those resulting in the achievement of goals $<G_k>$ such as (INCREASE (STUDENT SCORE)), say, and those resulting in the deterioration of progress towards goals $<B>_\ell$, such as (INCREASE (COMPUTER TIME)), say. A measure μ relating to the likelihood of the overall achievement of goals can then be computed for each L_i . for example

$$\mu = \sum_{k=1}^{n} W_k \times P(G_k) - \sum_{\ell=1}^{m} W_\ell \times P(B_\ell)$$

where $P(M_j)$ is computed as $\beta \uparrow \mathrm{POSS}(M_j)$ $(0 < \beta < 1)$. Where $\mathrm{POSS}(M_j)$ is as above the number of causal assertions involving the POSS operator used in the deduction of M_j from $<L_i>$, where W_j is the weight associated with the goal variable which is the argument of the predicate M_j .

7. The list of executable predicates, $<L_i>$ associated with maximum value of μ is then selected.

8. The arguments of the executable predicates, $<L_i>$, are then tested against a list of primitive arguments. These primitive arguments are the names of the variables of the state-vector and the names of the teaching operations which may be found on the right-hand side of the condition-action rules. Where a primitive argument is not found the definitional assertions are used to instantiate one. For example, if the executable predicate is (RAISE THRESHOLD) but no variable of the state-vector is called THRESHOLD then a definitional assertion such as (DEFEQUIV THRESHOLD (OR TECHNIQUETHRESHOLD PROBLEMTHRESHOLD)) can be employed to give (RAISE TECHNIQUETHRESHOLD), (assuming TECHNIQUETHRESHOLD is the name of an element of the

state-vector).

9. The set of executable predicates can then be executed.

A detailed example of the use of this deduction procedure is given in chapter 5.

Heuristics

In the above deduction procedure a number of heuristics are employed in comparing alternative sets of actions. In step 4 the number α of actions considered would be set according to the computational resources necessary to perform deductions (which depends, of course, on the size and complexity of the set of assertions). The measure μ defined and used in steps 6 and 7, and the value of β, are heuristics designed to select the sets of actions which are likely to result in the achievement of goals. For a given assertion (POSS ACTION EFFECT), and a given set of production rules, there may well be a probability value which could be associated with the occurrence of the EFFECT given that ACTION is executed. However, in the above formulation, it is assumed that this probability is not known. But we must choose between actions with possible or certain 'good' or 'bad' effects. Also, from consideration of the potential applicability of a probabilistic interpretation to an assertion of the form (POSS ACTION EFFECT), it seems desirable that 'possibly possibly' be weaker than 'possibly'. Accordingly (POSS ACTION EFFECT) has been arbitrarily assigned the probability β $(0 < \beta < 1)$ in the evaluation of μ.

An alternative approach would be to assign an individual probability p (for CERT $p = 1$, for POSS $0 < p < 1$) to each causal assertion; this probability being based on some *a priori* rough estimate. Such probabilities could be revised, where possible (that is, within limitations of generalisation capability), on the basis of the experiments carried out by the system. In this case $p(M_j)$ in the evaluation of μ would equal $\prod_{j=1}^{n}$ POSS(M_j) where POSS(M_j) is the probability value of the jth deduction where n deductions were employed to deduce M_j from L_i. This new definition of POSS would also be employed in step 4 (except now, of course, the L_i with the largest value would be the most 'likely').

But even allowing for the difficulty in making the initial estimate assigning probabilities to assertions would be problematic for two reasons. First, the probability will vary with the set of production rules the action is executed on. Second, the magnitude of the effect associated with the action is not known. With respect to this deduction procedure, the 'probability' should really be a combined estimate of the likelihood of the effect occurring and of its potential magnitude.

<u>Illustration</u>

Consider the following example sets of assertions where *Ai indicates an executable predicate Gi a predicate on a goal variable with a 'good' effect, and Bi a predicate on a goal variable with a 'bad' effect:

```
1.  (POSS A1 G1)
2.  (POSS A2 (OR G1 G2))
3.  (POSS *A3 (OR G1 B1))
4.  (POSS (AND *A4 A5) A1)
5.  (CERT (AND *A6 *A7) A5)
6.  (CERT A5 A2)
7.  (POSS *A8 (OR A2 B1))
8.  (POSS (AND A5 *A6) B1)
9.  (CERT (AND *A4 *A6) B2)
```

Then taking G1 as goal, steps 1 to 3 of the deduction procedure are illustrated by (a) of figure 6. Four possible courses of action L_i with associated values of POSS(L_i) can be identified, namely, L_1 = (AND *A4 *A6 *A7), POSS(L_1) = 2; L_2 = (AND *A6 *A7), POSS(L_2) = 1; L_3 = *A8, POSS(L_3) = 2; L_4 = *A3, POSS(L_4) = 1. Taking $\alpha = 2$ the possible courses of action for G1 are (AND *A6 *A7) and *A3 (step 4). The deduction of the possible consequences is given in (b) and (c) of figure 6. Assuming weights $W_i = 1$ and taking $\beta = 0.75$, then for (AND *A6 *A7):

$$\mu = p(\text{G1}) + p(\text{G2}) - p(\text{B1}) = 0.75 + 0.75 - 0.75 = 0.75$$

and for *A3

$$\mu = p(\text{G1}) - p(\text{B1}) = 0.75 - 0.75 = 0\,,$$

thus making (AND *A6 *A7) the preferred course of action.

Had actual probabilities been used, for example, if assertions 2, 6, and 8 were rewritten:

2. (PROBABILITY = 0.5 A2 (OR G1 G2))
6. (PROBABILITY = 0.3 A5 A2)
8. (PROBABILITY = 0.8 (AND A5 *A6) B1),

then for L_2 = (AND *A6 *A7), POSS(L_2) = 0.15 (with respect to G1 in step 4) and $\mu = 0.15 + 0.15 - 0.8 = -0.5$ (step 6).

Efficiency

The deduction procedure is liable to be very costly (both in processor time and store allocation) as it executes exhaustive searches of trees whose starting nodes are either a goal or a set of actions. These trees can usefully be 'pruned' as follows. In the case of a start node which is a goal, only the α 'most likely' nodes at any point in the deduction procedure need be expanded. In the above example, taking $\beta = 0.75$, the tree (a) in figure 6 is pruned to the tree (d). The deduction procedure then takes on the character of a

Figure 6: Example of the operation of the deduction procedure.

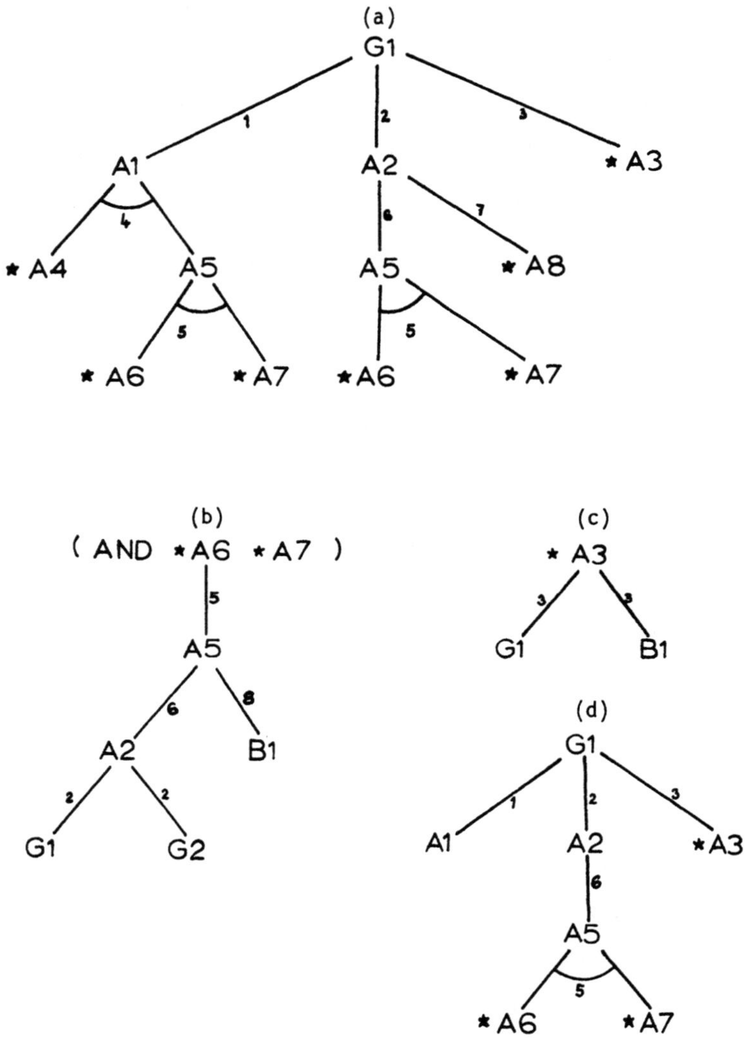

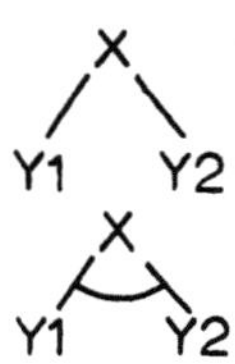

i - indicates assertion *i* employed.

— indicates Y1 OR Y2 will achieve X.

— indicates Y1 AND Y2 will achieve X.

directed search of a tree guided by the function POSS. In the case where the starting node is a set of actions, a branch can be terminated if the 'probability' of a consequence is below some threshold γ ($\gamma < 1$), say 0.05.

Using such techniques the computational cost of the deduction procedure will depend on the values chosen for α, β (if employed) and γ. A comprehensive survey of tree-searching techniques is given in Nilsson, 1971.

An evaluation procedure

The procedure above, in effect, generates hypotheses of the form "executing changes (AND $C_1 \ldots . C_n$) on the tutorial strategy may possibly assist in achieving a desired change in the value of a goal variable G_i". The next step is to execute the changes, evaluate them, and create new assertions as the result of the evaluation. Note that some new assertion must be created or the next time goal G_i is selected the same set of changes may be deduced as a possible course of action. The evaluation procedure terminates and creates new assertions if any of the following conditions are fulfilled:

a) It proves impossible to execute the changes (AND $C_1 \ldots C_n$). For example, it would not be possible to execute (REMOVE OPERATION1) if OPERATION1 were absent from the tutorial strategy. In such a situation the assertion (POSS* (AND $C_1 \ldots C_n$) IMPOSSIBLE) is created.

b) If as a result of the change in tutorial strategy the ensuring teaching performance is no longer 'acceptable' (see 2.5), for example, the program continues indefinitely asking the same question, then again the assertion (POSS (AND $C_1 \ldots C_n$) IMPOSSIBLE) is created and the

---

* CERT might seem more natural, but see discussion later (3.5).

tutorial strategy prior to the execution of the change is restored.

c) If a statistically significant (at some preset level Z1) change in teaching performance with respect to any of the goal variables G_i has occurred. If, for example G_j has increased, the assertion (POSS (AND $C_1 \ldots C_n$) (INCREASE G_j)) can be created. In addition, a measure of overall improvement

$$OI = \sum_{i=1}^{m} W_i G_i \quad (W_i = \text{weight on goal variable } G_i)$$

is computed. If OI has increased the new tutorial strategy is adopted, otherwise the old strategy is restored.

d) If a statistically significant decrease (at some preset level Z2 , $Z2 \leqslant Z1$) occurs in OI the previous strategy is restored and the assertion (POSS (AND $C_1 \ldots C_n$) (DECREASE OI)) is created.

The above evaluation procedure embodies the experimental method of the self-improving program. The assumption is made that sets of actions are independent. The procedure is only really workable in terms of simple statistical tests (e.g. t-tests) on individual goal variables. These and other associated limitations are discussed in the next section (3.5). However, 'running experiments', even of this limited sort, make it possible to try out and then discard or assimilate alternative courses of action. One question which arises is the extent to which it is useful and possible to generalise experimental results.

<u>Generalising experimental results</u>

A causal assertion created as a result of experiment will directly relate executable actions to changes in the value of goal variables. Such an assertion will be implicitly generalised by any definitional assertions operating on any elements of the clauses of the new assertion. For example, if the assertion (CERT (RAISE THRESHOLD1) (INCREASE STUDENT-TIME)) is created then this would be generalised by definitional assertions such as (DEFEQUIV THRESHOLD (OR THRESCHOLD1 THRESHOLD2 THRESHOLD3)) or (DEFPOSS STUDENT-TIME (OR TEST-TIME HELP-TIME)). Because of the way the deduction procedure is defined, causal assertions such as (CERT (RAISE THRESHOLD) (INCREASE STUDENT-TIME)) and (POSS (RAISE THRESHOLD2)(INCREASE TEST-TIME)) would now be implicit.

In many cases it may be desirable to rewrite existing causal assertions in addition to creating new assertions. This can be done easily in the case where the experimental result directly contradicts an existing assertion, for example (POSS (RAISE THRESHOLD1) (DECREASE STUDENT-TIME)) should be deleted if (CERT (RAISE THRESHOLD1) (INCREASE STUDENT-TIME)) is created (assuming a definitional assertion of the form (DEFEQUIV DECREASE (NOT INCREASE))). Also in the case where a number of similar causal assertions are created, for example, (CERT OP_1 $EFFECT_1$), (CERT OP_2 $EFFECT_1$) ... (CERT OP_n $EFFECT_n$), a definitional assertion (DEFEQUIV OPA (OR OP_1 OP_2 ... OP_n) and a causal assertion (CERT OPA EFFECT) can be created to replace the above.

However, the problem of rewriting causal assertions in general is very difficult. Consider the three causal assertions

1. (POSS(RAISE THRESHOLD1) (KEEP BORED-STUDENTS))
2. (POSS (KEEP BORED-STUDENTS) (INCREASE GUESS-RATE))
3. (POSS (INCREASE GUESS-RATE) (DECREASE STUDENT-TIME))

Suppose in addition the clauses (KEEP BORED-STUDENTS) and (INCREASE GUESS-RATE) occur in other causal assertions. Now suppose the causal assertion (CERT (RAISE THRESHOLD1) (INCREASE STUDENT-TIME)) is created.

This contradicts the deduction (POSS(3)* (RAISE THRESHOLD1)(DECREASE STUDENT-TIME)) which can be made from (1) to (3). Clearly, (1), (2), or (3) cannot be deleted as the cause of the new causal assertion may be independent of the 'keeping of bored students' or 'increasing the guess-rate' (or may outweigh these considerations in this particular local situation). The only reasonable generalisation (with respect to the deduction procedure described) in the above example is to create the new 'weak' assertions (POSS(2) (KEEP BORED-STUDENTS)(INCREASE STUDENT-TIME)) and (POSS(3) (INCREASE GUESS-RATE) (INCREASE STUDENT-TIME)). Adopting such a procedure could result in a proliferation of weakly contradictory assertions. (Although very weak assertions with probability less than some δ , say, could be discarded.)

An approach more in tune with actual experimental practice in CAI would be to use possibly contradictory assertions as a basis for generating hypotheses and experiments about intermediate constructs. In the above example this might involve defining more exactly or redefining such constructs as (KEEP BORED-STUDENTS) or (INCREASE GUESS-RATE) and running experiments on the effect of this. This type of process would involve model-building and experimental design of a sophisticated sort and is quite beyond the scope of this study. It would also offer a means of avoiding the simplifying assumption that the effect of actions on the teaching strategy are independent.

Summary

At points in the discussion of the deduction procedure various alternative approaches were presented. In each case the choice was

---

* POSS(N) stands for 'possibly possibly possibly' 'N' times. POSS(1) is written POSS.

between efficiency and maximal utilisation of the theory of instruction expressed by the set of assertions. The appropriate choices will depend on the computational environment (computer and language), the complexity of the theory, and the premium placed on speed of self-improvement. A detailed example for one particular context is given in chapter 4. The alternatives are listed below.

a) The value chosen for α the number of alternative courses of action possibly leading to the desired goal which are to be compared.

b) The value chosen for β - the 'probability' of an effect following an action. Another alternative is to employ guesses of probability for each causal assertion and to use experimentally determined probabilities in the creation of new assertions.

c) The value chosen for γ in pruning, that is, when to terminate deductions leading to consequences of low probability.

d) The value chosen for δ in determining the scope of generalisations.

e) The way in which goals are selected. (Step 1 in deduction procedure.) It would be more efficient in terms of the deduction procedure to select a single goal and deduce and execute all of a number of different courses of action before updating the set of assertions.

f) The strength assigned to assertions created as a result of experiment. If executing ACTION results in EFFECT as determined by a statistical test with a confidence interval of $x\%$, which assertion should be created -- (i) (CERT ACTION EFFECT), (ii) (POSS ACTION EFFECT), or (iii) (probability = $x/100$ ACTION EFFECT)?

3.5 Critical analysis of deduction procedure

Limitations

There are a number of severe limitations to the deduction schema described. Consider the causal assertion (CAUSE X Y) which could be interpreted as "for the current set of production rules, executing change X will result in the occurrence of event Y ". In many cases, particularly when assertions are being created and added, assertions of this form would seem to be more natural and more useful than assertions of the form (POSS X Y). The difficulty arises because of a variant of what is known as the 'frame problem' (see McCarthy and Hayes, 1969). In our case, while (CAUSE X_1 Y) may be true for a given set of production rules, we have no way, apart from experiment, of determining if it is still true after the execution of, say, change X_2 . Accordingly, the weaker (POSS X_1 Y) is used.

However, in some instances, for example, if with the current set of production rules a change has a highly statistically significant effect on a variable, we wish to distinguish that change. In this circumstance an assertion of the form (CERT X Y) is added. This solution is unsatisfactory in that this use of CERT does not strictly conform to the interpretation of CERT given earlier.

A related difficulty arises with 'impossible' changes. In our current formulation, a change X_1 executed upon the set of rules may result in the impossibility of applying X_2. X_2 may then be tagged IMPOSSIBLE to prevent the deduction system from repeatedly deducing that this is the best candidate for execution. But a further change X_3 may result in X_2 now being possible to execute and in fact the best candidate. In our existing formulation it is not possible to remove the tag IMPOSSIBLE from X_2.

The program is also liable in principle to 'hill-climbing' problems in that at any time it always executes the most likely change. It may be the case, however, that the execution of some change with a small 'bad' effect will make possible the execution of a sequence of changes with very 'good' effects. The program may, therefore, never carry out some optimum sequence or combination of changes even though each particular change may be described in some causal assertion.

Whether this problem occurs will depend on the topology of the search space. Hill-climbing problems will only occur in practice if there are strong localised interactions between experimental changes. The problems discussed above thus stem from the assumption on which the experimental procedure is based, namely that experimental changes are independent of each other in their effect on the tutorial strategy.

An important limitation is the very restricted form of generalisation of experimental results discussed above in 3.4. This is of course a fundamental problem in any work on the mechanisation of learning (see Plotkin, 1971). In our particular application the problem of generating reasonable generalisations of experimental results is made more difficult by the fact that individual assertions are employed for one or more of the following functions:-

a) to indicate possible lines of experiment;
b) to indicate the methodology for carrying out experiments;
c) to express a theory of instruction.

Another difficulty is that although in principle the value of β for a given assertion may be high (≈ 1) or low (<0.1), this value (even if it can be estimated at a particular time by experiment) will vary with the changes executed on the tutorial strategy. This is another variant of the 'frame problem'. In fact in practice β was arbitrarily taken to be 0.75 (see chapter 5). Another difficulty is the interpretation of 'possibly possibly' - an assertion so qualified is taken to have 'strength' $(0.75)^2$.

While some aspects of this formulation such as the CERT and POSS operators are adapted from modal logic (see Hughes and Cresswell, 1972), the formulation cannot be identified with any of the recognised modal calculi. It is essentially a 'common-sense' approach to the problem of reasoning about actions, restricted to the particular context of selecting changes to be executed on a set of production rules embodying teaching strategies.

Generality

In principle the approach presented could be employed to construct self-improving programs for any domain where:-

a) a strategy of actions can be expressed as a set of condition-driven production rules.

b) a theory of the possible utility of changing the actions can be expressed as a set of modally qualified propositional assertions.

It should be clear that in addition to the limitations discussed above the potential for self-improvement depends on the initial set of production rules and on the initial set of assertions in the theory. It should also be clear that although the assertions can be very general or highly specific, the theory as a whole will represent not very much more than an economically expressed and partially ordered list or agenda of possibly useful experiments.

CHAPTER 4

An implementation of a self-improving teaching program

In order to evaluate the design given in the second chapter and the techniques described in the third chapter, it was necessary to implement a self-improving teaching program for a non-trivial subject area. The design and techniques were developed and refined in the process of implementing the various incarnations of the teaching program. The original design remained largely unaltered, but the working of the deduction schema had to be amended (the distinction between causal and definitional assertions and the various techniques used to promote efficiency were not originally envisioned). Also, during implementation a number of limitations became clearer and better understood (see, in particular, 3.5 and 6.3).

This chapter begins with a discussion of why the chosen teaching domain was particularly appropriate for this investigation. A preliminary experiment was carried out in which various possible teaching programs were mimicked. This experiment and the transformation of its results into the teaching component of the program are described. In this component the tutorial strategy and student models are expressed as sets of ordered production rules. A description of the particular sets of production rules implemented is followed by an account of the experimenting component of the self-improving program, which is designed to execute experimental changes on the production rules. This component employs a theory of instruction for the teaching domain which is also described. The last section of the chapter lists various implementation details and various unexpected dificulties that had to be overcome in the course of implementing the program. This section concludes by identifying the task-independent features of the implementation.

4.1 The teaching domain

Ideally it would have been desirable to test and develop the ideas presented in the previous chapters by constructing a number of programs for different teaching areas and by testing them with large numbers of students. In the time available this was not possible. Accordingly it was necessary to identify a teaching task for which it would be comparatively easy to implement the teaching component of the system. This task should lend itself to interactive teaching (as opposed to non-interactive exposition), and should not make too great a demand on the time of potential students. However, for the domain to be amenable to the evaluation of the self-improving component, the task should elicit a variety of possible student performances. This would ensure that the teaching component would have to adapt to individual student learning characteristics. The teaching task should also have no obvious optimal teaching strategy associated with it.

It proved surprisingly difficult to identify a suitable teaching task that would be practical to implement as a teaching program. Eventually it was suggested* that the task of teaching the solution of quadratic equations by the discovery method would be an appropriate and rigorous test of the design. This suggestion was posed as a challenge. The simplicity of the teaching operations involved is more than outweighed by the complexity of the various possible teaching strategies, and the severe difficulty in determining any changes in

---

* by Professor R.C. Anderson, University of Illinois, Visiting Fulbright Fellow, Computer Based Learning Project, University of Leeds 1971-72.

student competence during a teaching session.

It should be emphasised that this subject area and teaching style provide a test bed for the ideas in the previous chapters. No attempt is made in this thesis to systematically evaluate or champion the discovery method as a teaching style either in general or for this particular subject matter. For information, a number of comments on its efficacy are offered in the next chapter (see 5.3).

The rationale for the discovery method style of teaching is that, if the student himself discovers solutions and methods of solution, this will deepen his understanding, promote his retention of the material and maintain his motivation. For a general discussion of the discovery method, see Schulman and Keisler (1966).

For quadratic equations, a teaching session using the discovery method style of teaching usually proceeds as follows. The student is presented with equations of the form:

$$\square^2 + c = b \times \square$$

and asked to guess the solutions, (x, y). Initially values of b and c are chosen such that the solutions are easy to guess and are easily seen to be true, for example, $c = 2$, $b = 3$, yields $x = 1$, $y = 2$. The problems are then made more difficult and the student is asked to guess solutions. The student is judged to have discovered how to solve this class of equations when by applying appropriate rules he is able to correctly state both solutions. The rules he is supposed to discover are that $x + y = b$, ("ADD" rule) and $x \times y = c$, ("TIMES" rule). Subsidiary rules which he may discover during the session and which can assist in the solution of the problem are that both x and y are factors of c, ("DIVIDE" rule), and that if $b = c + 1$, then $x = 1$ and $y = c$, ("ONE" rule).

The teaching strategy centres on giving the student carefully chosen examples which increase the likelihood of a student discovering a particular rule. Eventually, the student will appear to have mastered and be applying a particular rule. He is then presented with examples for which his rule is not sufficient, interspersed with examples for which his rule is adequate. This represents an attempt to discourage him from rejecting his rule, while at the same time encouraging him to try to discover other rules.

Student performance varies greatly, but most students eventually start developing hypotheses about ways in which the problems can be solved and testing them out. Whether the student finally succeeds depends primarily on his arithmetic skills and his ability to attend to the problem and systematically test hypotheses.

A number of factors contribute to making this a fairly demanding teaching task. The students may through chance have a fair degree of success while applying a completely spurious rule. Students discover, lose, and rediscover rules with great frequency. Students reject correct rules as a result of incorrect arithmetic or faulty logic.

The range of ability is also very great. Some students 'see' the rules almost instantly. Others need to be exposed to twenty examples before they develop an idea of what sort of rule is needed to solve the equations.

Consideration of these factors indicates that for a teaching program to be effective it should adapt to individual students. Also, the optimal teaching strategy is not at all apparent. Therefore it seems very appropriate to implement an adaptive self-improving program for this subject and teaching style.

The next section describes in greater detail possible teaching strategies and the way in which student performance varies.

4.2 A preliminary experiment

A preliminary experiment was carried out in a primary school in Leeds with twenty ten-year old pupils.

The experiment

First the pupils were given an introductory session in which it was determined whether they could solve equations of the form:

$$\square + 3 = 12$$

and

$$\square^2 = 9$$

or a word problem isomorph of these equations; for example, "If a boy has a bag of marbles and adds three marbles to the bag and then tips out the bag and finds twelve marbles, how many marbles were in the bag initially?". In some cases a little practice and explanation was necessary, but eventually all the pupils could easily solve these pre-test equations.

The pupils were then presented with a series of quadratic equations in the style given above or in a rather clumsy word problem isomorph (see Appendix I), and encouraged to guess at solutions. The pre-tests and teaching session were conducted through the medium of a set of cards with instructions written on them. When the experimenter felt that none of the cards in the current set were adequate for a situation, a new card was created. This approach proved to be a reasonable way of evolving a repertoire of teaching operations which could be applied in a teaching program. The use of stereotyped responses on cards also facilitated the recording of protocols.

Discussion of protocols

Prior to the experiment, the class teachers were convinced that it would not be possible for pupils in this class, except possibly the brightest, to solve quadratic equations, let alone discover a method of solving them. In fact, the majority of the pupils eventually discovered the method of solution without explicit prompting in periods ranging from half an hour to two hours. All the pupils, except one (who thought the exercise was silly), obviously enjoyed guessing and were clearly very pleased with themselves when they isolated the rules. Many of the pupils, having discovered rules, kept on demanding harder examples for them to demonstrate their new skill on.

During the sessions the pupils' motivation was affected by a number of factors. Some pupils had a very strong dislike of making mistakes and being told that a guess was wrong. They would guess rarely and reluctantly. This was resolved by emphasising the game-like aspect of the task and by substituting "yes" and "no" for "right" and "wrong". When a pupil discovered one of the rules there would be a big upsurge in interest. Occasionally a pupil would become dispirited and start guessing lots of high numbers or in a series:- 10, 11, 12, 13, ..., etc. In this situation it seemed best to give him the solution and proceed with another 'easier' example.

A number of features were observed to contribute to task difficulty for pupils who had not discovered the rules for solution. Most of the pupils had a fair idea of the reasonable range of values for their guesses. So increasing the value of b or c increased the difficulty. Holding b or c constant over successive examples made it easier for pupils to pick up the "ADD" rule or the "TIMES" rule respectively. Where the factors of c were 2 and a prime (> 5), this seemed to assist the acquisition of the "TIMES" or

"DIVIDE" rule. For some reason that is not apparent to me, if b was a largish prime (> 7) this appeared to assist the acquisition of the "ADD" rule. Some pupils, having already acquired the "ONE" rule, attempted to apply it to every equation.

Pupil performance and strategies varied considerably. If they acquired a rule early in the session (particularly the "ONE" rule), they tended to apply it frequently. If they acquired a rule later on they would discard it easily. Some pupils were very systematic and would construct and test hypotheses. Others were very unorganised in their approach. It was surprising to discover that some of the pupils had a strong self-image of themselves as being lucky. Having guessed correctly a couple of times, they regarded it as a waste of time to look for rules that would help them, as they knew they could guess the right answers. Other pupils were also adept at finding patterns (of the type normally employed in I.Q. tests) in the succession of solutions to the equations. Most pupils, having formulated a rule, would test it on previously given examples. A small number of pupils would persistently adhere to a rule despite obvious contradictions. One boy appeared to be making deliberate errors in his arithmetic to fit his incorrect rules.

Pupils would often find a first solution by guessing, either randomly or with the aid of the "DIVIDE" rule. The "DIVIDE" rule could be used to restrict guesses to factors of c . If they knew the "ADD" or "TIMES" rule, they could then apply one of them to the first solution to obtain the second.

After the pupils had mastered the rules they were asked to articulate them and, with one exception, could do this. The exception, who could solve difficult quadratic equations quickly, banged his head and pronounced: "It's in there, but I can't make it come out!"

Of course none of the pupils understood the problem in terms of the secondary school formulation; that is,

Let $x^2 - b \times x + c = 0$ then if x_1 and x_2 are solutions

$$(x - x_1) \times (x - x_2) = 0$$

$$x^2 - (x_1 \times x) - (x_2 \times x) + (-x_2) + (-x_1) = 0$$

$$x^2 - (x_1 + x_2) \times x + x_1 \times x_2 = 0$$

$$\therefore \quad c = x_1 \times x_2 , \qquad b = x_1 + x_2$$

A month later I tested some of the pupils, and they were still able to successfully apply the rules and solve quadratics. In fact, this exercise seemed to have given the pupils a lot of confidence in their approach to their school mathematics.

Implications for the teaching component

The above discussion has a number of very clear implications for the teaching component of the program. The pre-test was discovered to be more important for settling the pupil down than for weeding out pupils unable to cope with the task. The program should be friendly ("No, have another go"), rather than stern ("Wrong"). Also, the program should intervene if a pupil has clearly given up (that is, lots of random, unlikely guesses).

The features which were judged to affect task difficulty are the size of b and c, the number of factors of c, whether b is prime, whether a solution is "1", and whether b or c are held constant over successive examples.

An important implication for the construction of the student model is that student guesses do not precisely reflect the student's state of knowledge. For example, a gap often occurs between getting the first solution and the second solution, even when the student has acquired a rule which should allow him to get the second solution directly after the first. Accordingly, the model will have to make estimates based on the student answers as to whether he has acquired or is testing a particular rule.

With respect to the tutorial strategy, it became apparent that the program must be very conservative, and in general should give students who appear to have acquired a rule a great deal of reinforcement (that is, lots of examples easy to solve with that rule). However, an excess of such reinforcement may make it harder for the pupil to eventually realise that his rule is useful but not adequate to solve every equation. This is the type of trade-off problem which must be resolved by the self-improving component of the program.

Accordingly, the tutorial strategy must either contain or be easily amended to contain production rules based on such features of the teaching task. So that, for example, there should be production rules setting the level of 'reinforcement' for particular problem solutions and classes of students. Similarly, it will be necessary to have assertions in the theory of instruction which relate possible changes in levels of reinforcement to predicted changes in student performance.

4.3 The teaching component of the program

This section describes in detail the teaching component of the self-improving program. First the pre-test administrator is described. Then the four components of the teaching program proper, which are derived from Hartley's (1973) structure for CAI programs, namely:-

vocabulary of teaching operations (i.e. task administration), representation of task (i.e. task selection), model of the student (i.e. hypothesis tester), and means-end guidance rules (i.e. tutorial strategy). Although designed and implemented to run with the experimenting component, it could be, and was on occasion, run independently as a non-self-improving teaching program.

The pre-test administrator

This is a small adaptive testing program. The program uses a simple task difficulty measure (number of 'carries') to solve various simple linear equations, and a simple adaptive presentation strategy expressed in production rules. These rules provide a good example of how production rules can be used to express simply and transparently an adaptive strategy. This set is given in its entirety in Appendix VIII. This program tests whether the student has the basic arithmetic and algebraic skills to master the quadratic task. Students who solve the pre-test problems easily are quickly given a couple of criterion problems and the program passes control to the teaching program proper. Students who find the problems hard are given a sequence of examples which slowly increase in difficulty and some explanation of their mistakes. This process also yields a rough measure of student ability, namely the number of mistakes made. Some protocols of the pre-test program are given in Appendix I.

Task administration

This involves administering examples and general book-keeping. Administering examples includes presenting the example, giving encouragement ("well done", "have another go", etc.), and responding to guesses ("nope", "that was a wild guess sunshine", etc.). These functions are best illustrated by the protocol in Appendix I. Book-keeping functions include keeping track of the order and type of examples presented, and updating the state-vectors of the student model, and the state-vector of the tutorial strategy.

The working and inter-relation of the various parts of the teaching component are illustrated in figure 7.

Example selection

The model of task difficulty derived from the preliminary experiment was expressed in terms of a task difficulty matrix which relates teaching goals and subgoals (for example, that the student should acquire the "ONE" rule) to example features. These features are weighted as to how much they help or hinder the acquisition of the various subgoals. The task difficulty matrix is given in Appendix V.

The example selection procedure takes an input command of the form (GEN ((RULE1, HELP), (RULE2, HIN), PROPERTY)). This is interpreted as "generate an example which will help the student acquire or apply RULE1, which will hinder the application of RULE2 and for which PROPERTY is true". Using the task difficulty matrix, the example features are ordered with respect to their relation to the subgoals given in the command.

The features are then sorted according to two criteria. First, incompatible features are eliminated from the list. An example of a pair of incompatible features is "that one of the solutions to the problem be 1" and "that none of the solutions be 1". Then currently impossible features are eliminated. Consider the pair of features "that both solutions have numerical values less than 5". After a number of problems have been presented it may be impossible to find a new example with these features. The table of examples is then searched for the best fit to the remaining set of features. If no fit is found, the set of features is added to a list of sets of features now impossible to obtain. Then the feature with the smallest relative weight is

Figure 7: Schematic diagram of the teaching component of the quadratic tutor.

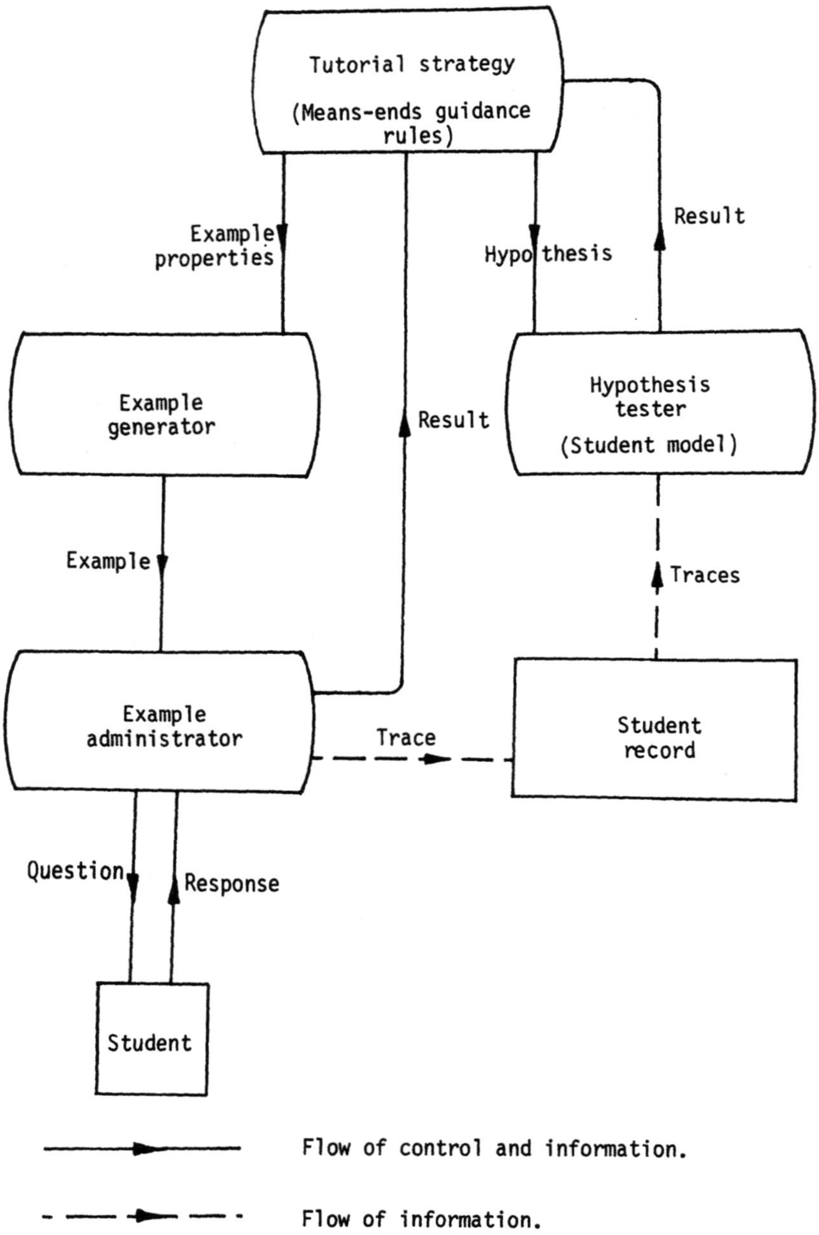

Flow of control and information.

Flow of information.

dropped and the search repeated until an example is finally selected and administered.

<u>Hypothesis tester</u>

While some assumptions about student learning are implicit in the tutorial strategy, estimates of the student's current state of knowledge are made by a hypothesis tester. This constitutes the explicit part of the student model.

The five hypotheses that may be tested are as follows:

1) WORULE - that the student has no rule at all.
2) WRULE - that the student has completely mastered all the rules.
3) TIMES
4) ADD
5) ONE

(3–5) - that the student has mastered the particular rule in question.

As the discussion of the preliminary experiments indicates, evaluating these hypotheses presents certain difficulties. In particular, students often will appear to have completely mastered a rule and yet fail to apply it in appropriate situations, with the result that there is an occasional 'gap' (that is, one or two incorrect guesses) between the first and second correct solutions. Also, phenomena such as temporary forgetting, lucky correct guesses and correct guesses which could have been achieved by more than one rule, make it very difficult for the hypothesis tester to definitely accept or reject a hypothesis. The hypothesis tester takes as input factors such as the size of the problem space (that is, the current number of legal guesses) and previous estimates of the student's state of knowledge, along with the student's guesses as he solves the current equation.

This difficulty in confirming or rejecting hypotheses is partly resolved by returning one of a range of values: CERTAIN, VERY POSSIBLE, POSSIBLE, DON'T KNOW, POSSIBLY NOT, CERTAINLY NOT.

Each of the five possible hypotheses has a set of production rules associated with it. These express a simple, explicit decision procedure for testing each hypothesis. The elements of the state-vectors include the number of guesses, the 'gap' between the correct solutions, the result of any previous evaluation of this hypotheses, and a measure of the size of the problem space (value of c in $x^2 + bx - c = 0$). Further, each state-vector has variables indicating the occurrence of particular key features in the student responses related to the hypothesis being evaluated. For example, the state-vector for the TIMES rule includes a variable NUMFACT: the proportion of guesses for the last example which are factors of c. The complete set of rules for evaluating the hypothesis that the student has the TIMES rule is given in Appendix IV.

Tutorial strategy

In one sense the tutorial strategy may be said to encompass all the decisions taken during the implementation of a teaching program: decisions such as the format of the questions, the entries in the task difficulty matrix, the nature of the student model, and the order of examples generated. What is described in this subsection is the set of means-end guidance rules which relate the predictions from the student model to the process of task selection. These rules constitute a presentation strategy for the different example types, and co-ordinate the other components of the program, namely the task administrator, task selector and student model (again see Figure 7).

The presentation strategy is expressed as a set of production rules. The set is given in Appendix III. The 19 elements of the state-vector are variables associated with tutorial strategy such as the current teaching subgoal, the 'reason' for the last tutorial action executed and estimates of the student's state of knowledge from the student model. In a complete cycle of operation the existing state-vector is first parsed using the partitions (for detailed worked examples see chapter 3 or Appendix III). The left-hand sides of the rules are then searched for the first match with the parsed vector. The corresponding right-hand side, which in fact is a list of LISP function calls (see C. Weissman, 1967), is then executed and the state-vector is updated as a result.

The following functions are used on the right-hand side.

a) (SETQ NAME X) - used to set parameters such as GUESSLIM: the maximum number of incorrect guesses permitted on any one example.

b) (HYP RULE) - test the hypothesis that the student has mastered the rule in question (see section on Hypothesis Tester).

c) (GEN ((RULE1, HELP), (RULE2, HIN)) (PROPERTY)) - select, and administer example with appropriate features.

d) (CONTINUE N) - generate and administer 'N' more examples with the same properties as the previous one.

e) (GOAL NAME) - sets a flag in the state-vector indicating the current subgoal.

f) (DPRINTER NAME) - output the text labelled NAME.

The production rules can mostly be divided into those relating to the testing of hypotheses and those concerned with example selection. An example of the former is: ((1 A3 19 K1) ((HYP ONE))). This reads "if the program is in the hypothesis testing cycle and the current subgoal is the ONE rule, only test whether the student has acquired this rule". The rationale behind this rule is that testing the other hypotheses would be a waste of computer time. However, continual checking of all hypotheses would result in the program detecting incidental learning of other rules sooner. This again is the kind of trade-off problem which the experimenting component will try to resolve.

The rules relating to example selection may be considered as describing a partially ordered set of goals. The goals fall into four classes:

a) ADD, TIMES and ONE - get the student to master the rule named.

b) ONETOADD, ONETOTIMES, ADDTOONE - get the student to master the second named rule without losing the first named rule.

c) EASY - give the student problems solvable by inspection.

d) HARD - give the student examples only solvable with complete mastery of the task.

The interrelation of the goals is illustrated by the directed graph given in Figure 8. Each node represents a goal, and the outgoing arcs correspond to the production rules associated with the goal. For example, GOAL TIMES may be left if conditions expressed by the right-hand sides of any of the rules labelled T1 through T6 are met (see Appendix III). The conditions these correspond to are:

a) T1 - "has the student mastered and successfully applied both the ADD and TIMES rule for the last three examples?"

b) T2 - "has the student mastered the TIMES rule but not the ADD rule?"

c) T3 - "is incidental learning on the ADD rule greater than learning on the TIMES rule?"

d) T4 - "is it 'very possible' or 'certain' that the student has mastered the TIMES rule?"

e) T5, T6 - check for incidental learning on ADD and ONE rules respectively.

Figure 8: Directed graph showing interrelation of tutorial goals and subgoals of quadratic tutor.

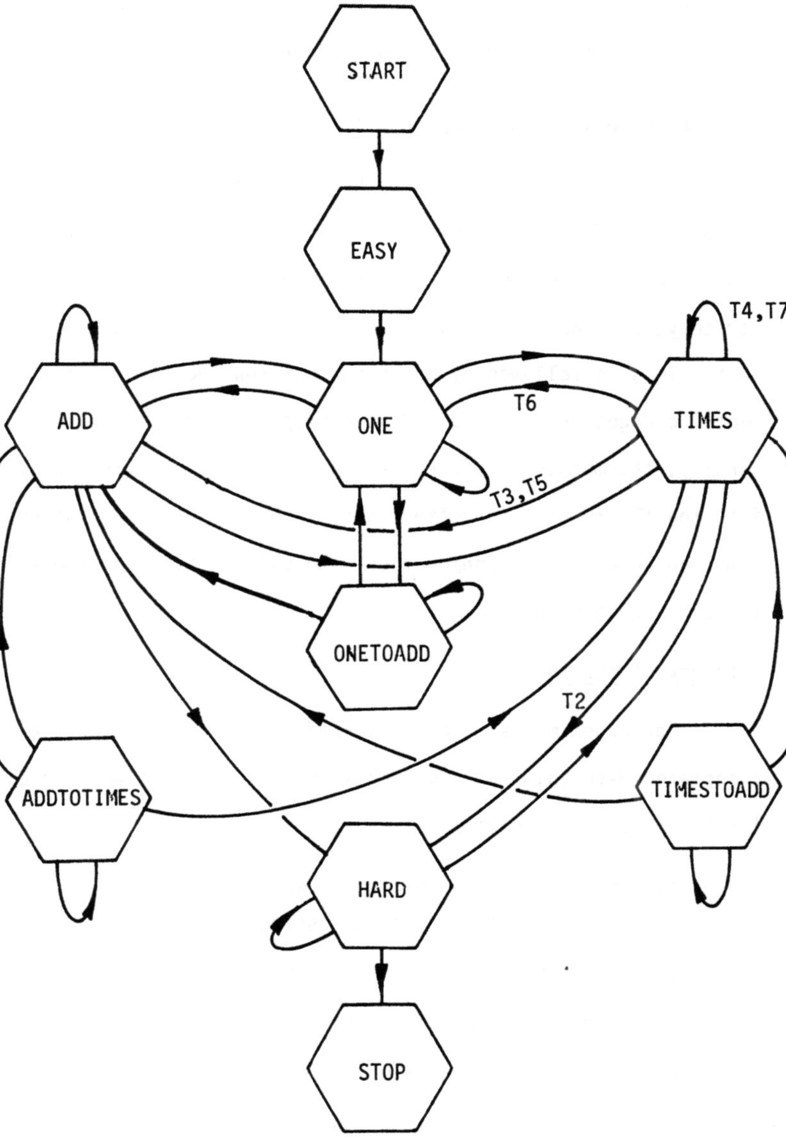

In the event of none of these conditions being fulfilled, rule T7 insures that example selection for this goal continues.

The complete set of production rules expresses the tutorial strategy in an explicit fashion amenable to automatic manipulation by the experimenting component of the program.

4.4 The experimenting component of the program

This section describes a theory of instruction for the teaching domain and teaching component given above. The deduction procedure which is employed to derive possible changes in the tutorial strategy, and the amender which executes these changes, are given next. This is followed by an account of how the effect of experimental changes on teaching and student performance is evaluated. Finally, the method of incorporating the result of this evaluation into the theory of instruction is described. The way in which this component operates in conjunction with the teaching component is illustrated in figure 9.

A theory of instruction

For experiments to be evaluated, overall goals and measures associated with them must be identified. Four goals were selected for the program, namely:

a) Increase the number of students successfully completing the teaching session, (SS).

b) Increase the average score on a post-test, (PS).

c) Decrease the amount of student time used, (ST).

d) Decrease the amount of computer time used, (CT).

The theory of instruction consists of assertions relating general or highly specific changes in the tutorial strategy to the

Figure 9: Schematic diagram of operation of self-improving quadratic tutor.

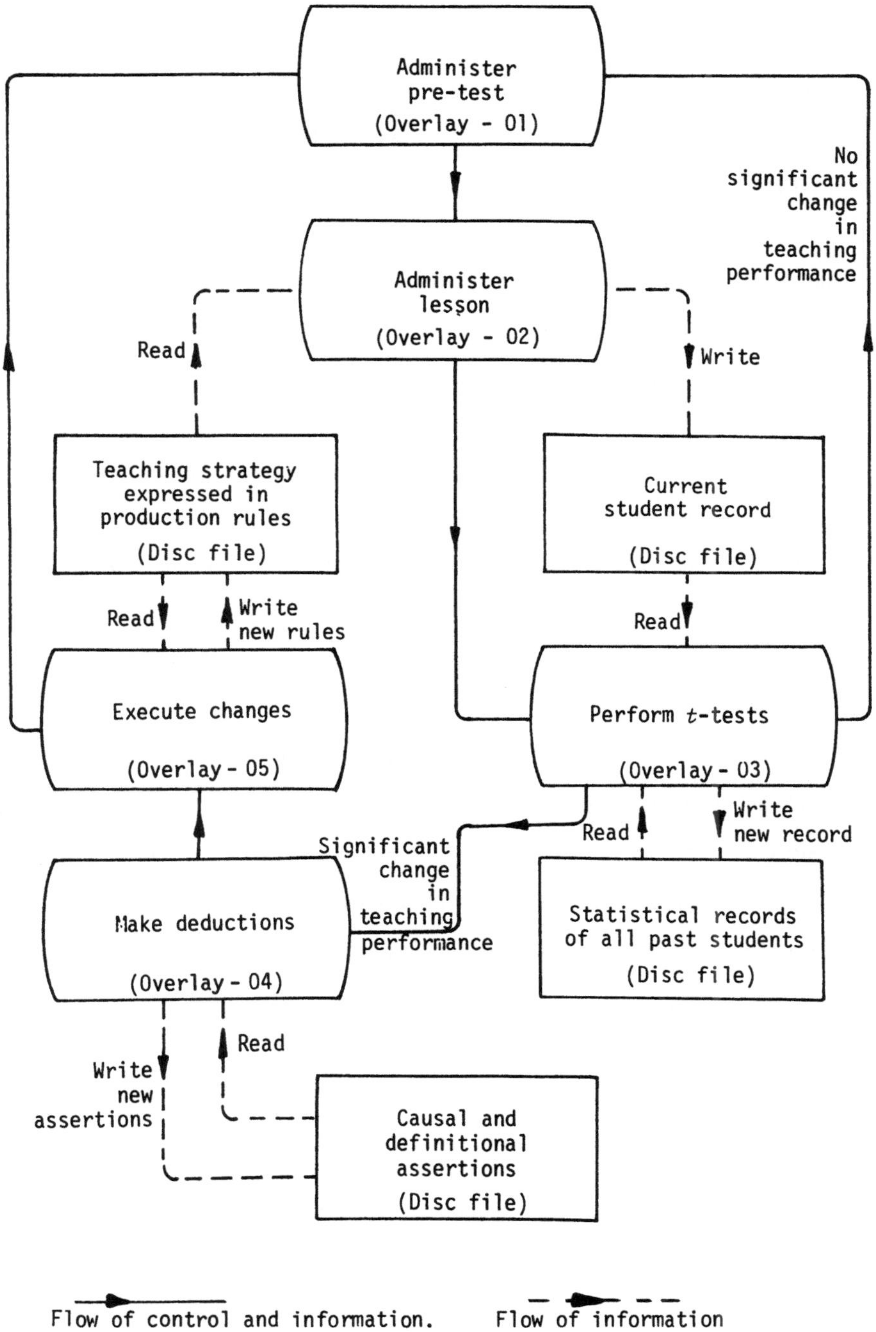

goals above. It would be possible to create such assertions about changes in the sets of production rules which constitute the student model. This was not done as it seemed unlikely that it would even be possible (because of time constraints) to examine all the possible changes on the tutorial strategy. The theory of instruction which was used for the experiments described in the next chapter is given in full in Appendix VI.

The syntax of the assertions is as given in 3.3. Some of the causal assertions are highly specific to the teaching component given above. For example, some refer to possible changes in parameters set in the tutorial strategy, e.g., (POSS (LOWER GUESSLIM) (DECREASE (STUDENT TIME))). This can be interpreted as "if the set of production rules constituting the tutorial strategy are changed so that the variable associated with task administration which sets a limit on the number of guesses per equation is lowered, then a decrease in student time may possibly result". (A list of parameters associated with task administration which may be reset by the tutorial strategy is given in Appendix III, section e.)

Other assertions refer to changes in the order of specific teaching subgoals, for example, (POSS (AND (AFTER (GOAL EASY)) (REPLACE (GOAL ONE) (GOAL TIMES))) (SHORTEN SESSIONS)). Other assertions are very general and could be reasonably found in any self-improving program with the same goals; for example,

```
(CERT (SHORTEN SESSIONS) (DECREASE TIMES))
(POSS (SHORTEN SESSIONS) (DECREASE SCORES))


(POSS (DECREASE (HYPOTHESIS TESTING))
      (OR (DECREASE (COMPUTER TIME))
          (INCREASE SCORES)
          (INCREASE (STUDENT TIME)) )   )
```

Note that some of the definitional assertions are also highly specific, e.g. (EQUIV (OR REPLIM SYSLIM GUESSLIM WILDLIM) EXLIMIT). In this case the class of parameters which assert the number of guesses permetted is being defined. This sort of definitional assertion is also of use in the amender (see 4.5).)

Other definitional assertions are completely genetal, e.g. (EQUIV CHANGE (OR INCREASE DECREASE)). All the specific assertions must clearly be written with reference to the existing set of production rules (or any form it might be amended to), and name elements which occur or might occur in it. General assertions are or no use if the constructs they employ are not related by other assertions to changes in goal variables and changes in tutorial strategies. For example, the causal assertion: (CERT (KEEP BORED STUDENTS) (INCREASE TIMES)) would never be applied without assertions indicating how (KEEP BORED STUDENTS) might be achieved by changes in the tutorial strategy.

The deduction procedure

This is derived from the deduction procedure in chapter 3.4. Some detailed examples of deductions on the above theory of instruction are given in chapter 5.

β, the arbitrary 'probability' assigned to possible assertions, was set at 0.75. For a given goal, α, the maximum number of alternative possible courses of action was set at 2. This deduction incorporated all the tree-pruning techniques for increasing efficiency that were discussed in 3.4 except that γ was set $= 0$. So only two nodes were expanded at any stage in the deduction process but very unlikely deductions were not discarded.

The amender

The amending primitives implemented are those described in 3.2, with the exception of INCLUDES, EXCLUDES, NARROW, WIDEN, ADD and DELETE. The amender operates as described in 3.2. Some examples of amending which occurred during the course of experiments are given in the following chapter.

The amender operated with an index listing the nineteen elements of the state-vector, identifying the elements which measured progress towards the tutorial subgoals, listing the primitive argument (e.g., ≡MULT, ≡EASY) and executable predicates (e.g. REMOVE, INSERT, etc.) and associating tutorial actions with goals. The latter was achieved simply by associating any action of the form (HYP ≡X) or (GEN ≡((X HELP) ....)) with the goal X . Actions not associated with any particular goal such as DPRINER, CONTINUE, HYPALL were indexed with all goals.

Statistical evaluation

A simple statistical program was implemented for the purposes of the experiment. It maintained records of all the students with respect to the four goal variables. The program carries out t-tests after each additional student has been taught. These t-tests are used to detect significant changes in the mean scores subsequent to the last experimental change in the set of production rules. If there is a significant change, using a 90% confidence interval, in any of the goal variables, the change is then evaluated to determine if it should be permanently incorporated in the set of production rules.

OI, the measure of overall improvement, was computed as a linear function of SS (student score), PS (post-test score), ST (student time) and CT (computer time). ST and CT were given negative coefficients, and the weights were chosen so that for the units in which SS, PS, ST and CT were measured a 5% increase in SS or PS score was 'worth' roughly 20 additional student minutes at the terminal or 10 seconds of processor time. Otherwise the set of production rules prior to the change is restored. Small sample statistics and evaluation after each student were necessary in view of the small number of students on which it was expected to run the program.

Modifying the theory of instruction

A program which was based on the proposals presented in 3.4 for automatically modifying theories of instruction was implemented. This program creates causal assertions whenever a statistically significant relation is established during the evaluation of an experimental change to the tutorial strategy. It also creates causal assertions by applying all appropriate definitional assertions. These new assertions are 'weak' (that is, modified by POSS operator) generalisations of the main result. Assertions may always be added or deleted by the experimenter if he wishes to give the program some 'advice' or to 'prime' it in some way. Examples of this program in practice are given in the next chapter.

4.5 Implementation notes

The final version of the program was implemented in LISP and ran interactively under Taurus, a time-sharing system on the CDC 6600/6400 at the University of Texas at Austin.

The system comprised five overlays each occupying 33K and four disc resident files. The operation of the system is illustrated in Figure 9 on page 103. Wherever possible the LISP code had been compiled (see Greenawalt, 1974) for speed. An attempt was made to minimise the amount of evaluation in executing the production rules. For example, parsed symbolic state-vectors were stored, and the parsing only recomputed for those elements of the state-vector which had been altered since the last parse of the state-vector. An entire lesson used on average about 50 seconds of processor time. When the deduction procedure and amender were used, a further 60 to 90 seconds of processor time were used.

The response time of the teaching component ranged from 2 to 5 seconds when the program was run between 8:00 a.m. and 10:00 a.m. Occasionally, teaching sessions were held later in the day, when there were many more users on the time-sharing system. The response time could then be as poor as one to two minutes.

In praise of LISP

The first version of this program was implemented in BCPL (see M. Richards, 1969). It proved much easier to implement an effective version in LISP. But, of course, the demands on processor time and space (core and disc) were considerably higher.

The most important advantage of LISP is the ease with which it is possible to pass code or calls to functions from one LISP program to another. It added greatly to ease of implementation, clarity, and elegance to have the facility for detaching the ACTION side of a production rule or to deduce a possible change in the deduction procedure and then promptly execute that action or change without any intermediate level of interpretation.

Equivalent implementations are of course feasible in BCPL, but necessitate the construction of various interpreters. The task administrator and amender were originally implemented in BCPL as two separate interpreters. The other important advantage for this type of programming was the facility with which LISP programs can be changed by other LISP programs. For example, each set of production

Features which changed during implementation

A number of features envisioned in the original design (see O'Shea and Sleeman, 1973) had to be changed or abandoned for pragmatic reasons.

In an early version of the program, a non-deterministic select function (see Floyd, 1967) was used to generate examples. While elegant, this proved to be tremendously inefficient, and table look-up was found to yield equally effective examples at the cost of considerably less computer time.

The generalisation scheme employed in practice was simply the creation of weak generalisations of new experimentally determined assertions. The statistical evaluation program was never extended beyond *t*-tests in view of the small number of students available.

However, a number of techniques were successfully evolved to overcome some difficulties that had not been anticipated at the design stage. The distinction between causal and definitional assertions was created to avoid infinite chains of deduction. Some tree-pruning techniques were also evolved to increase the efficiency of the deduction procedure.

Another innovation was augmenting the set of production rules with goal 'markers' or annotations in order to make possible the directed graph interpretation of sequences of teaching subgoals. A particular element of the state-vector was used to mark this. The amender was extended to incorporate the primitives BEFORE, AFTER, etc.

for changing the order in which goals were selected. This greatly simplified the problem of representing tutorial strategies and amending them. The original design given in chapter 2 was thus largely achieved

Task-dependent and -independent features of the program

This chapter has concentrated largely on the features of the program directly related to the teaching domain. A number of the procedures are, however, task-independent. The teaching component uses a general production rule handler for the tutorial strategy and student models. Also, a fairly general procedure for inputting and outputting formatted problems and recording responses is utilised by the pre-test and task administrators. All the features of the experimenting component, apart from the values of α , β , and γ employed in the deduction procedure (see 3.4) are task-independent.

The amender, however, utilises an index to the components of the tutorial strategy. This index would have to be set up appropriately to use the amender with a tutorial strategy for another task domain.

The important features which are completely task-dependent are: a) the formatted problem text used by the task administrator; b) the task difficulty matrix; c) the various sets of production rules; and d) many of the assertions in the theory of instruction.

So, the implementation of a self-improving program for a similar (i.e. algebraic) subject area would not involve an inordinately large amount of reprogramming.

The task administrator is limited to quadratic equations, and an alternative task administrator would have to be rewritten in its entirety. Compared with most CAI programs the system described here is very modular and has expressed in a completely explicit form the tutorial strategy, hypothesis tester, task difficulty matrix and example generator. Such components, if present in conventional CAI programs, are commonly diffusely implemented, and changes to them necessitate substantial reprogramming. Here the entire tutorial strategy is expressed as six sets of production rules and a matrix. So any reprogramming of the tutorial strategy is limited to formulating new production rules or computing new elements for the matrix.

CHAPTER 5

Experiments with the Program

One of the usual ways of experimenting with a teaching program is to monitor the program's teaching performance with a view to ironing out any undesirable features which become apparent. Another complementary approach is to carry out experiments on aspects of the teaching strategy and employ the results to refine the teaching strategy.

Experiments of the latter type were carried out by the self-improving program itself. Both types of experiment and the teaching performance of the program are discussed in this chapter. (See also O'Shea, 1973.)

5.1 Experimenting with CAI programs

Performance monitoring

This type of experiment is routinely carried out on newly implemented CAI programs. Usually it consists of monitoring the performance of the teaching program from the typical user's point of view to ensure that the program is sufficiently 'friendly'. The notion of 'friendliness' covers aspects of performance such as:- response time; how comprehensible the questions the program asks the student are; how easy it is for the student to ask for help; how useful the program's prompts are; the overall response-sensitivity of the tutorial strategy (for example, are all students consistently asked such easy questions that they are bored, or are the questions so difficult that they are confused and discouraged?); and lastly, is there any unpredicted teaching performance as a result of undetected 'bugs' in the program?

This performance monitoring phase is described in section 2 of this chapter. Some of the problems identified in this phase can be corrected by simple measures such as changing the text of questions or prompts or by inserting 'traps' in the example administrator (for example to check whether the student has typed in "help"). Other problems will be identified for which there is no simple solution. For example, if a proportion of the students are clearly bored by being stepped through a set of questions which is too easy for them then it is not appropriate to just increase the level of difficulty of the problems. It will not be apparent from the performance monitoring phase how large the increment in example difficulty should be. Further, there may very well be a trade-off problem. For example increasing the example difficulty may result in a proportion of students becoming hopelessly lost. In such a case if the program is to be improved it will be necessary to carry out an experiment.

Experimenting and refining

Some experiments will be on the effect of simple changes in the tutorial strategy (such as changing part of the sequence of presentation) on teaching performance. Other experiments will be concerned with refining the student model or task difficulty model and changing the tutorial strategy to take advantage of this refinement to deal with the problem concerned. The program must then be subjected to experiments of the performance-monitoring type again. For as well as checking that the problem has in fact been overcome, it is also necessary to check that there is no unpredicted interaction between the modified parts of the teaching program which results in deterioration of teaching performance. Good examples of this type of experiment are the series of experiments associated with the Leeds arithmetic program (see discussion in 1.2).

This thesis is concerned with ways of automating this process of successively refining a teaching program by carrying out experiments on aspects of teaching strategy. Apart from an initial monitoring phase to check for intelligibility of instructions and for bugs in the program, the subsequent refining and monitoring experiments were carried out by the program itself. Altogether, five refinements of the teaching strategy were experimentally carried out by the program, and these experiments are discussed at length in this chapter. These experiments are used to demonstrate the practical feasibility of the design and techniques presented in this thesis. They are also used to illustrate some of the limitations inherent in this approach. Some points at which a human experimenter can usefully intervene in this self-improving process are indicated.

5.2 Initial monitoring of program performance

The first experiment was carried out with four fourteen year-old students at Austin High School. The students were observed while using the teaching program on an interactive terminal at the school. After completing the lesson with the program they were requested to indicate any sources of confusion which had arisen during the lesson. They were also asked for any other comments on the program. From this experiment it was possible to make the following general conclusions about the program's teaching performance:

a) Students could learn how to solve quadratic equations as a result of using the program.

b) The response time of the program (two to five seconds) was adequate to maintain interest and did not provoke any severe lapses in attention.

c) The students enjoyed using the program.

The following particular changes in the teaching program were indicated and subsequently executed:

a) The students had a lot of trouble with the first few quadratic equations encountered. Two easy quadratics were incorporated into the pre-test with demonstrations that their associated solutions were correct.

b) Students were a little unsure of what they were doing while guessing values to the qudratics. The introduction of the prompt "LOOK FOR RULES" appeared to improve matters.

c) The students had been told that there would be two distinct solutions to each problem. But when guessing the solution to a quadratic equation they sometimes repeated a guess which had already been responded to as 'correct'. Then they were confused when after getting two correct responses the example was not terminated. An appropriate prompt was introduced - "YES BUT YOU HAVE GUESSED THAT CORRECTLY BEFORE".

d) Students often forgot to press the return key on the console and a reminder was included in the introduction to the pre-test.

e) Some questions were obscured by the position of the teletype printer after printing. This was remedied by printing two line-feed characters after each question or prompt.

f) Some students typed their responses as "x=7" while others typed just the number "7". The example administrator was amended to accept both forms of response.

g) The prompt "FIND THE VALUES" was found to be confusing and was replaced by "GUESS THE VALUES".

h) Students were confused when asked to type their 'first name' (apparently this is not an American expression), and this was changed to 'Christian name'.

Most of the observations and the remedies listed above would have been hard to anticipate in advance. Consider, for example, the last observation (h); it would be difficult to anticipate and pointless to incorporate the potential for such changes into a self-improving program.

There were a number of observations to which no immediate action obviously corresponded. These observations included the following:

a) Pre-test performance appeared to be a good predictor of performance during the main teaching sequence.

b) Early on during the main teaching sequence, comments such as "TRY AGAIN" or "THAT WAS A WILD GUESS SUNSHINE" were very important and had a large effect on student interest and rate of working. After a dozen examples they appeared to have a slight negative effect.

c) The prediction of the student model seemed too conservative. Students appeared certain they had a rule two or three examples before the hypothesis tester was 'certain' they had acquired the rule in question.

The objection to making changes in the program on the basis of the above three observations is that they were based on a very small number of students and that in each case the exact type of change that should be made is not apparent. However, these are the type of observations which may lend themselves to formulation as assertions and incorporation into the 'theory of instruction'. The assertions may then subsequently initiate relevant experiments during the self-improving process. For example, the causal assertion (taken from Appendix VI)

```
(POSS(LOWER HYPTIME) (OR(DECREASE TIMES) (DECREASE
                                         (POST SCORE))))
```

incorporates observation (c).

A more general discussion of how new assertions may be created by observers of the program is given in 5.5.

5.3 The program in use

After the initial monitoring of program performance described in the previous section the program was used by fifty-one students. The students were aged between thirteen and fifteen years. Although these students did not belong to the age group (nine to eleven years) for which the teaching program had been designed, they could all be identified as having difficulty with mathematics. The students had all failed at least one high school mathematics course and were attending remedial mathematics classes. None of the students was able to solve quadratic equations at the start of the lesson.

The lesson was presented to the students on an interactive terminal located at the school. The lessons lasted on average about an hour, during which time the student attempted an average of 24 problems. The day after the lesson the students completed a written post-test comprising six problems to be completed within ten minutes.

A typical protocol from a student using the program is given in Appendix II. This sequence of problems was generated by the set of production rules constituting the tutorial strategy given in Appendix III. In this protocol the first goal adopted by the tutorial strategy is to get the student to acquire the 'ONE' rule. The student appears to have acquired this rule after the second problem has been administered. The conservatism of the hypothesis tester and the production rule governed transition to another goal (which generated two problems after the hypothesis tester was 'certain' the student had the rule), resulted in four more 'ONE'-rule problems being administered before the goal 'ONETOADD' was adopted. (See 4.3 for a discussion of the various goals.) The seventh problem (marked [P7] in the protocol) is directed to the 'ADD' rule, and the student fails to solve the eighth problem which is again solvable by the 'ONE' rule. (This possibly justifies the program's earlier conservatism.) After the tenth problem, the goal 'ONETOADD' is dropped, since the student has shown himself capable of identifying and applying the 'ONE' rule where appropriate. Subsequently going through the program's adoption of the 'ADD', 'MULT' and finally 'HARD' goals there are sequences of problems where it seems that the program is being too conservative, followed by instances where it suddenly seems that maybe the program is being too optimistic. Consider the respective sequences leading to the fifteenth and to the twenty-second problem. There is a lot of scope for experimenting with the production rules that make the tutorial strategy conservative or optimistic over wide or narrow classes of tutorial situations. Similarly there is potential for experimenting with the sequence of tutorial subgoals and the various thresholds such as the maximum number of systematic guesses permitted.

Some comments on the discovery method

There are reasonable grounds for criticising the use of this teaching style in this context. It presents as a guessing game a topic which students could be more thoroughly, and perhaps more quickly taught (students, that is, with the appropriate background) by discussing the appropriate algebraic derivation of the solution.

No emphasis is placed on the underlying mathematical principles or on very practical interpretation of quadratic equations. It would seem unwise to base entire mathematics syllabuses on the learning of problem-solving heuristics by guessing rather than on the learning of the skills, methods and habits of mind of mathematical analysis.

However, it must be noted that this criticism does not take into account the motivational effects of 'discovery' solutions. Before using the program most of the students characterised themselves as very poor at mathematics and expressed very low confidence in their ability to learn mathematics. 74% of the students discovered all the coefficient rules for solving the equations, and could solve by inspection equations such as $x^2 + 48 = x \times 19$ after completing the lesson. The scores on the post-test were mostly high, the mean score being 81%. Most of the students enjoyed using the program and were very enthusiastic about both using the terminal and the teaching style of the program. A selection of student comments is given in Appendix VII.

Experimental procedure

The teaching program, without the self-improving component, was used to give lessons to the first twenty students. The students' records so collected provided the basis for evaluating the effects of subsequent changes in the teaching strategy on teaching performance.

The self-improving component was initialised as follows:

i) The magnitude of changes executed by the (RAISE X) and (LOWER X) function calls was set at 50% of the current value of X .

ii) The confidence interval for the t-tests was set at 90%.

iii) The maximum interval between changes in the teaching strategy was set at seven lessons. So if no statistically significant change had occurred after seven students had been run a new change was selected and executed. The previous change was to be incorporated if OI (the measure of overall improvement) had increased, otherwise it was to be removed.

Had a much larger number of students been available it would have been possible to select a smaller magnitude of change in (i), a greater confidence interval in (ii), and dispense with (iii). For the purposes of the experiment (i) - (iii) were set to insure that the self-improving program made a reasonable number of changes without being completely erratic.

After the program was initialised for the running of experiments it was employed to administer lessons to a further thirty-one students. The program operated as described by figure 9, with control initially at overlay 04, the deduction procedure. To illustrate the procedure employed, the cycle of operations used in the first experiment is described here.

The first goal the program adopted was to decrease the amount of student time. The deduction procedure described in 3.4 was employed to deduce the best possible lists of changes in teaching strategy that might achieve this end and would possibly result in minimal deterioration of performance on the other goals. These changes were then executed by 05, the production rule amender, on the set of production rules embodying the tutorial strategy. The modified set of rules were then employed by 01 and 02, the pre-test administrator and tutorial strategy, to administer lessons to students. After each student completed his lesson a t-test was carried out by 03, the statistical evaluator, to test for a significant change in teaching performance. This continued until seven lessons had been given. The change was incorporated into the teaching strategy as OI (the measure of overall improvement) had increased and control passed back to 04, the deduction procedure, and the cycle of deduction, modification, statistical check continued for four further experiments. The five experiments are discussed in detail in the next section.

5.4 The system's experiments

The system carried out the following five experiments, selecting goals in turn from a 'round robin' list.

1. The first goal adopted was to decrease the amount of time spent by students using the program, i.e. (DECREASE (STUDENT TIME)). (REMOVE MINOR) was deduced to be the course of action possibly facilitating this goal with minimum deterioration with respect to the other goals. As MINOR is not a 'primitive argument' but defined as the name of a 'class of goals' (see chapter 3), the appropriate definitional assertion was applied to give (REMOVE (GOAL ≡EASY)). This was then passed to the amender and executed.

The production rules that were identified to be changed were:

```
a)  ((2 B1)((DPRINER START1) (GOAL ≡EASY)
            (GEN ≡((EASY HELP)) NIL) (CONTINUE 1) ))

b) ((19 K9)((GOAL ≡ONE) (GEN ≡((ONE HELD) NIL)
            (CONTINUE 1) ))
```

(a) was identified by the "(GOAL ≡EASY)" on the right-hand side. (b) was identified by "K9", a partition associated with (GOAL ≡EASY). In rule (a) the REMOVE function deleted the action (GOAL ≡EASY) and the actions listed after it, (GEN ≡((EASY HELP)) NIL) and (CONTINUE 1), which are both associated with (GOAL ≡EASY) in the amender's index. The list of actions from the right-hand side of rule (b) was appended to the list of actions remaining in rule (a), and rule (b) was deleted. The net result of executing (REMOVE (GOAL ≡EASY)) was to leave the following rule in the position of rule (a):

```
((2 B1)((DPRINER START1) (GOAL ≡ONE)
       (GEN ≡((ONE HELP)) NIL) (CONTINUE 1) ))
```

The result of this change was that the teaching program no longer presented two easy introductory problems. The changes in the mean score of goal variables was monitored after each additional student was taught. After the seventh student had been taught using the amended tutorial strategy, no significant (using the t-test with a 90% confidence interval) change in any of the goal variables had been detected. In fact the mean value of student time had increased. However, computer time had decreased and post-test scores had increased. The value of OI (overall improvement) had increased, so the change in the tutorial strategy was kept. The inconclusive result of this experiment illustrates the difficulties of running simple experiments on single variables where in fact the factors being examined for change are not independent. The changes in the various scores over the whole series of experiments are summarised in the Table of Scores.

2. The goal of increasing the post-test score , (INCREASE (POST SCORE)) was next adopted. The deduced change in tutorial strategy was (REPLACE HYPS (HYPALL)), where HYPALL is the function which tests all the possible hypotheses. HYPS is not a primitive argument and is replaced by applying the definitional assertion

```
((AND (HYP ≡ONE) (HYP ≡ADD)(HYP ≡TIMES) (HYP ≡WRULE)
      (HYPALL)) HYPS) .
```

The result is that the six production rules with instances of one or more of (HYP ≡ONE), (HYP ≡ADD) or (HYP ≡TIMES) have these instances replaced by (HYPALL).

This change affected teaching performance in that the program now tested all its hypotheses in situations where previously only the hypotheses directly related to a tutorial goal had been tested. Thus

Table of scores of goal variables from experiments with the quadratic tutor

	Comparison of means between	Result of Experiment		Student time - ST (minutes)	Computer time - CT (secs. of CPU time)	Student score SS - (%)	Post-test score PS - (%)	Overall Improvement OI
Initial Group (A) n = 20	-	-	M =	51.9	52.9	75	83	53.2
			SD =	15	8.85	44.4	22.1	
Experiment 1 Group (B) n = 7	A(n=20) B(n=7)	Inconclusive	M =	54.1	49.6	71.4	85.7	53.4
			SD =	5.98	6.43	48.8	29.9	
Experiment 2 Group (C) n = 7	A&B(n=27) C(n=7)	Success t = 1.99 (CT) t = 1.76 (PS)	M =	43.8	43.9*	71.4	100*	83.7
			SD =	7.62	8.33	48.8	0	
Experiment 3 Group (D) n = 6	A&B&C(n=34) D(n=6)	Fail t = 2.73 (PS)	M =	44.3	45.6	50	53.3**	13.3
			SD =	4.25	14.5	54.8	51.6	
Experiment 4 Group (E) n = 4	A&B&C(n=34) E(n=4)	Success t = 2.68 (CT)	M =	39.3	38.5**	75	75	72.3
			SD =	2.5	8.58	50	37.9	
Experiment 5 Group (F) n = 6	A&B&C&E(n=38) F(n=6)	Success t = 1.85 (CT)	M =	47.3	42 *	83.3	90	84
			SD =	10	9.25	40.8	16.7	
Groups (E) & (F) n = 10	A(n=20) E&F(n=10)	t = 3.61 (CT)	M =	44.1	40.6***	80	84	79.3
			SD =	8.67	8.68	42.2	20.6	

M - Group mean
SD - Standard deviation

Notes:
* - significant at the 10% level (two tail test).
** - significant at the 5% level (two tail test).
*** - significant at the 1% level (two tail test).

the program was experimenting with a trade-off between an increase in computer time against a possibly earlier detection of the occurrences of incidental learning.

After seven students had been taught a significant increase in mean post-test score and a significant decrease in mean computer time were detected. The former change possibly indicates that the quality of example selection had improved. The latter effect is accounted for by a large drop in the average number of examples administered. Another contributing factor is that the examples selected were possibly more appropriate, The assertions

(POSS (REPLACE HYPS (HYPALL)) (INCREASE (POST TEST)))

and

(POSS (REPLACE HYPS (HYPALL)) (DECREASE (COMPUTER TIME)))

were added to the set of assertions as a whole. As the value of OI had increased, the change in tutorial strategy was kept.

3. The next goal adopted was increasing the student score, i.e., (INCREASE (STUDENT SCORE)). The course of action deduced was (REPLACE HYPS (HYPALL)). However, as during the previous experiment exactly this change had already been executed it was not possible to execute this assertion, and the assertion (CERT (REPLACE HYPS (HYPALL)) IMPOSSIBLE) was created. In general in situations of this kind it would probably be more useful for the amender to delete assertions with the clause (REPLACE HYPS (HYPALL)). The deduction procedure was then invoked again and the action (RAISE ENCRATE) was executed. As ENCRATE did not occur on any action list, but was on the amender's index of changeable parameters, this was effected by directly increasing the parameter ENCRATE by 50%. If there had been an action (SETQ ENCRATE 2), say, this would have become (SETQ ENCRATE 3).

The effect of this change on teaching performance was that students got more prompts of the form, "have another go". As a result of a bug in the program which gave encouragement, the amount of encouragement was increased by much more than 50% (in fact about 300%). The students were peppered with encouragement and as a result started guessing with complete abandon. After six students a significant decrease in post-test score was detected. The assertion

```
(POSS (RAISE ENCRATE) (DECREASE (POST SCORE)))
```

was added to the set of assertions. As OI had decreased the previous version of tutorial strategy was restored.

4. With the goal of decreasing computer time (LOWER HYPTIME) was deduced, and hence (LOWER TIMESCH) as an appropriate course of action. TIMESCH is the eleventh element of the state-vector and indicates the time and direction of the last change in hypothesized ability with the TIMES rule. It was employed in three production rules:

```
a) ((1 A3 6 F1 11 G3 19 K3) ((HYPALL)) )

b) ((1 A2 6 F1 7 11 G3 12 G3) ((GOAL ≡HARD)  (GEN
          ≡((WRULE HIN)) NIL) ) )

c) ((1 A2 6 F1 11 G3 19 K3)((GOAL ≡TIMESTOADD)
     (GEN ≡((TIMES HELP) (ADD HELP)) NIL) )) .
```

The eleventh element in each of these production rules is G3. G3 is given by the partition (G3 GREATERP ELEMENT 2). The change was executed by creating the partition (G6 GREATERP ELEMENT 1) and by rewriting 11 G3 as 11 G6 in the three rules above. This change was equivalent to giving less practice with the TIMES rule after it was judged (by the hypothesis tester) to be 'certainly' mastered. After four students had used the program a significant decrease in computer time was detected. The two assertions

(POSS (LOWER HYPTIME) (DECREASE (COMPUTER TIME)))

and

(POSS (LOWER TIMESCH) (DECREASE (COMPUTER TIME)))

were added to the set of assertions. As OI had increased the change in tutorial strategy was kept.

5. Now with the goal of decreasing student time, i.e., (DECREASE (STUDENT TIME)), the action (LOWER HYPTIME) was again deduced. The assertion

```
(POSS (LOWER HYPTIME) (OR (DECREASE TIMES)
                          (DECREASE (POST SCORE))) )
```

associates (LOWER HYPTIME) directly with the goal (DECREASE (STUDENT TIME)). The new assertions created after experiment 4 associate (LOWER HYPTIME) more strongly with the desirable side-effect of (DECREASE (COMPUTER TIME)). The result is that (LOWER HYPTIME) had become the strongest candidate (that is, it facilitates the achievement of the selected goal and has best side-effects) for execution.

The changes executed were similar to those in 4 with instances of "12 G3" being replaced by "12 G6". The effect was again a significant decrease in computer time, and the new assertions created were

(POSS (LOWER HYPTIME) (DECREASE (STUDENT TIME)))

and

(POSS (LOWER TIMESCH) (DECREASE (STUDENT TIME))).

Summary

After being run with thirty-one students the system had carried out five experiments. One had no result, three had positive results, and one had a negative result. Eight assertions had been added to the program's set of causal assertions. Comparing the performance of the program over the first twenty students (the initialisation stage) and the last ten there was a decrease in mean computer time of 23%. This is significant at the 1% level using a two-tailed *t*-test. The mean student time, student score, and post-test score improved by 15%, 5% and 1% respectively. None of these changes is significant at the 5% level using a two-tailed *t*-test (see Table of Scores).

The improvements in computer time and student time were not at the expense of deteriorations in student score or post-test score which were near their maximum values at the start of the experiment. It is harder to detect significant changes in student time than in computer time. For example, the larger standard deviation of the student time of the initial group of students was contributed to by broken air-conditioning in the terminal room. This prompted visits to the soft-drinks machine at the other side of the school. This sort of difficulty might be overcome by monitoring the frequency of student activity at the console and basing any measure of student time on this. Strictly the *t*-test should not have been employed with the variable student score. As a result of the way this score was determined it was in fact a binary variable (any individual student scoring either 100% or 0%). A non-parametric test such as the χ^2 would have been appropriate.

5.5 Appraisal of program

In addition to the improvement in teaching performance the program's set of causal assertions had been added to. An incorrect assertion had been selected, evaluated and contradicted. Several other assertions incorporating the results of experiments had been created. As noted above, the fifth experiment was in fact partly carried out as a result of the assertions created after the fourth experiment. Given that the set of assertions had members of a general nature like:

(POSS (REPLACE MAIN MAIN) (CHANGE SOMEGOAL))

which reads "replacing some 'main' teaching operation by some other 'main' teaching operation may result in some effect on some goal", or

(CERT (SHORTEN SESSIONS) (DECREASE TIMES))

the program could have continued to execute experimental changes in the teaching strategy over a large number of students.

<u>Potential further experiments with the system</u>

If a large number of students had been available it would have been possible to refine the other major elements of the tutorial strategy (that is, the student model and the task difficulty model) as well as the means-ends guidance rules which were experimented with in this implementation. The hypothesis tester (the major part of the student model) consists essentially of five sets of production rules. In order to experiment with these it would be necessary to write assertions indicating the possible utility of the various measures of student performance employed such as GAP - the number of incorrect responses between the student's two correct guesses for a given problem. The existing production rule amender could be employed, though with a different index of primitive elements (see 3.2). It would comprise the range of certainty values, (≡CERT to ≡CERTNOT), and the various names for the elements of the state-vector. The sets of production rules for testing the various hypotheses are simpler than the set for the tutorial strategy. There are no goal markers, only single actions on the right-hand sides, and the partitions are simpler. So only a subset of the possible amending commands would be employed, namely LESS, GREATER, EQUAL, PRESENT, ABSENT, RAISE, LOWER, and REPLACE.

To experiment with the Task Difficulty Matrix it would be necessary to write a set of assertions concerned with task features and their possible relations to the various teaching goals (e.g. ONE, etc.). The subset of amender functions necessary to execute changes would be only RAISE and LOWER. These functions would have to be changed so that they operated on a group of elements in the matrix. Further, if the matrix is to be at all accurately 'tuned' it would be necessary to have the capability for indicating the degree of change in the matrix elements.

For both the additional classes of experiment discussed above, the existing system goals and method of evaluation would be appropriate. However, there are a number of weaknesses in the program which would have become apparent if large numbers of students had been taught. In particular, newly created assertions are not automatically assimilated by generalising them with appropriate existing assertions, and it is even not at all clear how an appropriate generalisation might be created in principle. So the set of assertions would continue to grow and become unwieldy and expensive in computer time to evaluate. Also, the system having tried out all its 'likely' changes, it would then start trying out the 'unlikely' ones. This would result in frequent short-term deteriorations in teaching performance.

In practice, if the system were used with any large number of students it would almost certainly be necessary for the system's experiments and the growth of the set of assertions to be monitored and for the observer to execute changes (i.e., 'non-automatic' changes) in the system at appropriate intervals. In particular the observer could create generalisations of groups of causal assertions by establishing appropriate definitional assertions.

Potential for non-automatic changes

With the system as implemented the observer would in practice probably be obliged to make house-keeping changes such as deleting redundant assertions and redundant production rules. However, the type of non-automatic change envisioned in the design of the program is the addition of assertions as 'advice'. Such advice would normally result from observations of the type discussed in 5.2. Consider

```
(POSS (LOWER HYPTIME) (OR (DECREASE TIMES) (DECREASE
                                            (POST SCORE))))
```

given in 5.2 as an example of an assertion to prompt experiments as to whether the student model is applied too conservatively. In this case it was possible to construct the causal assertion using predicates and arguments already in use such as LOWER, HYPTIME, DECREASE, etc. So it was only necessary to add one causal assertion to embody the advice.

However, in other cases it may be necessary to create definitional assertions defining the terms employed, and it may even be necessary in some cases to adapt the amender to the change. Consider the observation "pre-test performance appears to be a good predictor of performance during the main teaching sequence". One way of acting on this observation would be to create the causal assertion

```
(POSS (AND (CLEVER STUDENT) (LOWER (VAR GOALTIME)))
      (DECREASE TIMES) ).
```

This assertion indicates that possibly time can be saved by selecting fewer examples for each of the various teaching goals. Now (CLEVER STUDENT) must be defined. A definitional assertion for (CLEVER STUDENT) could take the form

```
(EQUIV (CLEVER STUDENT) (HIGH (VAR PRETEST))) .
```

In this case the amender must be adapted by making an entry in its table of primitives for (HIGH (VAR PRETEST)). The alternative is the inelegant

```
(EQUIV (CLEVER STUDENT) (PUT ≡QUADPARTS ≡PRETEST
                            (A1 GREATER ELEMENT 5))) .
```

Unfortunately cases like this will occur. Of course once dealt with they can be useful. In the above example any further cases of advice relating to students with high pre-test scores can employ the term "clever student".

It is worth noting at this point an important incidental advantage of a self-improving teaching system. As a result of the constraints inherent in constructing a program which can change itself, each component of the program is quite separate (i.e., modular). Further, the information on which the various different types of decision are made is of necessity completely explicit. This makes it easy to manually change any part of the system found wanting without excessive risk of introducing unwanted side-effects.

Evaluation of experimenting component

The results of the experiment should be approached with caution. Very small numbers of students were taught and a confidence interval of only 90% was used. A small number of experimental changes were carried out in practice. The only aspect of the teaching strategy experimented with was the tutorial strategy. However, the experiments do serve to illustrate how a self-improving program could operate in a practical CAI teaching role. To apply this system in other educational contexts or with other experimental designs it would almost certainly be necessary to extend the statistical evaluator so that it could perform other tests (such as analysis of variance and various non-parametric tests).

All the general criticisms of the approach given in 3.5 apply to this particular implementation. There is a hill-climbing problem in that the optimal sequence of experiments cannot be determined. There is a 'frame' problem; assertions are assumed to have a fixed probability of being true while in practice these probabilities will vary with the changes in the set of production rules. Another limitation is the assumption that the experimental changes are independent of each other. This results from the simple nature of the theories of instruction which can be expressed in the formalism put forward. In fact the formalism is no more than a fairly general and moderately elegant way of associating a set of changes that might be executed on a teaching strategy with some of the possible consequences of these changes. In contrast a more complex theory would be an attempt to account for the interaction of different changes on the teaching strategy and might have some associated meta-theory related to the alternative experimental design associated with the testing of the theory. The formalism presented here does not lend itself to the expression of complex theories of instruction, nor to designing experiments to test hypotheses generated from such a more complex theory.

The program was capable of improving its teaching performance but the scope for improvement depends on the initial set of production rules, and the limitations of the teaching style. The overall bounds to self-improvement depend on these and the initial set of assertions constituting the theory of instruction and the experimental methodology of single change followed by single experiment. In short, the behaviour of the experimenting component could be described as 'tuning' an existing tutorial strategy.

CHAPTER 6

Discussion and Conclusions

This chapter lists the contributions of the research. The range of application of the design is indicated and its practicality commented on. The chapter concludes with some comments on how the design could be extended and some notes on areas for further investigation.

6.1. Contributions

The contributions of this thesis are mostly to CAI. They take the form of illustrating how a number of techniques derived from work in AI can be applied to the construction of CAI programs with a self-improving capability.

Contributions to CAI

The principal contribution has been to demonstrate the feasibility of constructing an adaptive (that is, response-sensitive), self-improving teaching program. In particular, the following have been presented:

a) A general design for adaptive, self-improving teaching systems.

b) Examples of the use of production rules for the expression of aspects of tutorial strategy.

c) An amender for production rules which express tutorial strategies.

d) A simple formalism for expressing theories of instruction.

e) An associated deduction procedure for deducing experimental modifications to be carried out on the tutorial strategies.

f) An associated experimental procedure for carrying out and evaluating such modifications.

g) An implementation of a self-improving program which teaches the solution of quadratic equations by the discovery method.

The work described is particularly applicable to CAI programs intended to run under the very large multi-access systems such as PLATO (Dugdale and Kibbey, 1976) and TICCIT (Mitre, 1974) which are now becoming widespread. In such a computational environment decisions made in the construction of CAI programs affect very large numbers of students. This is combined with the possibility of collecting data on the performance of the teaching programs. With respect to such use, self-improving teaching programs have the following desirable characteristics:

a) They can run experiments and improve their own teaching performance.

b) They can collect teaching program performance data in a goal-directed and selective manner.

c) They have to be 'transparent' with respect to their tutorial strategy. That is to say, to allow for self-improvement, the various functions of the program which affect the tutorial strategy must be made explicit. (For example, in the design presented such functions are modular and are expressed as sets of production rules. Such transparency allows flexibility and other desirable side-effects.)

d) The overall educational objectives and any underlying theory of instruction must be made explicit.

Another contribution has been to provide further evidence for the utility of Hartley's (1973) proposed structure for adaptive teaching programs by successfully applying it to the design and implementation of the teaching component of the self-improving system.

Contributions to AI

Finally we claim to have made some small contributions to work in AI. Waterman's (1968) approach to the machine learning of heuristics has been extended and applied to a system which carries out experiments on a teaching program to test simple hypotheses about possible changes in tutorial strategy. These hypotheses are deduced by a deduction procedure which is a development of Black's (1968) work on question-answering. Taken as a whole the system is an example of a self-modifying program which is composed of many different autonomous but co-operating and interacting components - including a tutorial strategy, a student model, a task selector, a production-rule amender, a deduction procedure and a statistical evaluator.

6.2 Range of application of design

The design was developed with regard to potential application to CAI programs which have a teaching role and are response-sensitive. Subject areas which would lend themselves to this approach include high school mathematics, and subjects taught by drill and practice (for example, basic number skills, simple grammar, foreign language vocabulary). This includes the bulk of CAI work which adopts a generative approach, but would not include CAI programs without an explicit task analysis, for example author-language programs in which no distinction is made between statements which sequence teaching material and those which generate or select such material.

In principle the design given here could be applied to the construction of self-improving programs for any domain (that is, not just teaching) where the following can be separated and defined:

a) A set of explicit actions which can be associated with performance in this domain.

b) A strategy for executing these actions which can be represented parsimoniously in production rule form.

c) An algorithm for changing these strategies expressed as production rules.

d) A method for evaluating the effects of the actions.

e) A theory of the likely effects of the actions, or an agenda of possibly useful changes or experiments.

The two types of rule system employed in the self-improving teaching system are sets of simple ordered production rules (used for expressing tutorial strategies) and sets of modally qualified propositional assertions (used for expressing theories of instruction). If the design presented here were to be applied to some other non-trivial domain it would almost certainly be necessary to extend and develop these formalisms in order to adequately express features of that domain. Contemporary work with rule-based systems such as the Meta-DENDRAL program for mass spectroscopy (Buchanan and Mitchell, 1977) and the MYCIN program for antimicrobial medical consultation (Shortliffe et al., 1975) illustrate the variety of production rule formalisms which seem to be required.

6.3 Practicality of design

From a pragmatic computer science or CAI point of view the design presented has the two following good points. Firstly, it is modular. The theory of instruction, or the tutorial strategy, can be changed easily, either in part or as a whole. Secondly, the response time is comparable with that of a conventional CAI program. The deduction procedure and production-rule amender can be run 'off-line' when no students are using the program (and when demand for processor time is low).

A disadvantage inherent in the design is that the demand on some form of permanent memory (for example, disc or magnetic tape) is going to be greater than that for a conventional CAI program. A self-improving program must keep records of student and program performance, or experiments executed, and some past sets of production rules. However, the extra demand on store is unlikely to be excessive. For example, in the case of the quadratic teaching program, space was required for a spare set of production rules, the set of assertions, and 5 numerical scores per student taught. This occupied less than 5K of 60 bit words.

The risks associated with the design are listed below:

a) The amender might introduce a 'bug' in the tutorial strategy. For example, if an appropriate threshold were deleted the teaching component might get in a loop and teach indefinitely.

b) The quality of the hypotheses tested depends on the quality of the theory of instruction. If incorrect hypotheses are tested there will be short-term deterioration in teaching performance.

c) For very large numbers of teaching sessions the demand for permanent store for student records, etc. might become excessive.

To counter (c) it would be necessary to incorporate a routine to compact or selectively forget student records. To counter risks (a) and (b) it would be necessary to employ a critic for teaching performance. The critic would incorporate measures of acceptable teaching performance: for example, teaching must occur, more than three examples of a certain type must be given, less than 100 examples overall, etc. The critic could be expressed as a set of production rules associated with the same state-vector as the rules for the tutorial strategy. The partitions on the left-hand side of the rules would be employed to detect unacceptable teaching performance. It would, however, have only one permissible action for the right-hand side of the rule, namely -INTERRUPT, which has the effect of passing control to the deduction procedure. The deduction procedure should continue as if the experiment had failed and create new assertions accordingly. The rules constituting the critic would be applied after the PARSE of the PARSE, MATCH, ACT cycle of the tutorial strategy. (See Appendix III for cycle of operations.) It would be illegal to amend the set of production rules which made up the critic.

The design becomes much more practical if we assume that in practice self-improving teaching systems would be used as tools by teachers and experimental psychologists to develop and test teaching strategies and theories of instruction. In that circumstance if the teaching performance of the system greatly deteriorated the human experimenter could intervene and change any suspect production rules. Similarly if an interesting experiment or class of experiment became apparent the educationalist could add appropriate particular or general assertions to the set of assertions which would initiate these assertions. In this way a self-improving teaching system could be developed and extended in a similar synergistic mode to that successfully applied in MYCIN (Shortliffe et al., 1975). For the theory of instruction can be regarded as an action-driven production system (see Waterman, 1977) where the question to be answered by deductive inference on the assertions is "what change in teaching strategy will improve teaching performance with respect to educational objective x?".

6.4 Further research

The work described in this thesis is preliminary investigation into the design and viability of self-improving teaching systems. Further experiments that might usefully be carried out with the system as implemented and various possible extensions to the system are outlined below.

Further possible experiments

The self-improving teaching system described in chapter 4 is easy to experiment with provided sufficient numbers of students are available. Two obvious lines of experimentation with the system would be to examine the effects of changes in the initial tutorial strategy and in the theory of instruction on subsequent performance. In particular it would be of interest to determine just how robust the system is and under what circumstances its performance would converge. A related line of experimentation would be to alter the sequence of the various types of student taught by the system and analyse how this affected the modifications which the system carried out on its tutorial strategy.

It would, of course, also be useful to adjust the system for other subject areas in the manner described in 4.5. At present the system only makes changes to its tutorial strategy. But a large contribution to the response sensitivity of the teaching component is made by the hypothesis tester which constitutes the student model. Experimental changes could be executed on these by simply adding to the theory of instruction assertions which related properties of the student model to the educational goals of the system. In this way the student model might be refined as a result of teaching and experiment by the system.

Artificial students

One of the weak points of the program discussed in chapters 4 and 5 was the student model. This essentially consisted of a simple-minded hypothesis tester expressed as five sets of production rules. A useful approach might be to apply the results of the protocol analysis experiment described in 4.2 to the construction of a set of artificial students expressed in sets of production rules. An artificial student would be a program which interacted in the same way with the teaching program as some class of real student. Such artificial students could be employed in a number of useful roles in the context of a self-improving teaching program. Firstly, they could be used to replace the existing hypothesis tester in the following way. They could be 'run' on the same examples as the student being taught to find the artificial student which most closely matched the performance of the student. Secondly, they could be employed as a conjunct to the tutorial strategy. When a new example or piece of teaching material is being selected (or generated) it could be first tried out on the artificial student in order to predict its likely effect on the student's state of knowledge. Thirdly, artificial students could be employed in the role of a critic (in the sense of 6.3). They would provide a means of evaluating experimental changes in tutorial strategy for acceptability without subjecting real students to the experience of programs with 'break-downs' in teaching performance. For such a monitoring role it would probably be sufficient to employ two artificial students - one simulating the 'slowest' possible student, and the other simulating the 'fastest' student.

In the long term, by expressing artificial students as sets of production rules it may prove possible to achieve a synthesis between two areas of research related to the problem discussed in this thesis.

On the one hand there is the form of student model proposed by Self (1974) which would consist of a set of procedures to represent a student's current knowledge state. On the other hand, there is the work on the simulation of human problem-solving which employs production rules (for example, Newell and Simon, 1972; Waterman and Newell, 1973; Young, 1974). For example, Young (1977) has developed a series of models expressed in production rules for the various strategies which are employed in subtraction. These models could be developed into artificial students for a subtraction tutor. Changes in students' competence would be modelled by adding new production rules to the existing student model. Alternatively the approach used in Meta-DENDRAL (Buchanan and Mitchell, 1977) might be adapted to induce production rules to be incorporated into existing student models. This might provide a means of refining student models and thus increasing the response-sensitivity of the teaching system.

Theories of instruction

Another limiting factor in the current design is the formalism put forward for expressing theories of instruction. These have been discussed at length above. One limitation which could be partially overcome without radically changing the design is that although possibly true assertions may have different degrees of probability of being true they are arbitrarily assigned a probability of 0.75. They could be weighted more accurately with experimentally determined probabilities. However, the variant of the 'frame-problem' discussed in 3.5 will still apply.

Important difficulties remain. In particular the degree to which experimental results should be generalised varies with the result and its context (see 3.4). It is not at all apparent how to generate the most useful generalisation. Finally, the design proposed will neither support complex theories of instruction nor execute sophisticated experiments. To express more complex theories of instruction, so that for example possible interaction effects could be predicted or accounted for in the theory, it would be necessary to create 'meta-assertions' relating assertions to other assertions in the theory. For the development and testing of such theories it would be very important for the educationalist developing the theory to be able to easily change and add new assertions to the theories. Current work in rule-based intelligent programs which help experts (who are not computer scientists) to develop and extend them during use looks promising. See for example RITA, Anderson and Gillogly (1976) or TEIRESIAS, Davis (1977).

In conclusion, the system described here would be most usefully extended in two ways: firstly by using the system as a vehicle for developing and refining student models expressed as production rules; secondly, by incorporating some of the recent developments in rule-based systems (see Waterman and Hayes-Roth, In press) to human-engineer the system and to facilitate the expression and development of richer theories of instruction.

BIBLIOGRAPHY

Anderson, R.H. and Gillogly, J.J. 1976. Rand intelligent terminal agent (RITA): Design philosophy. Rand Report R-1809-ARPA, February.

Atkinson, R.C. 1972. Ingredients for a theory of instruction. American Psychologist, October, 921-931.

Banerji, R. 1974. Winston's theory of learning. SIGART Newsletter, April, page 5.

Baylor, G.W. and Gascon, J. 1974. An information processing theory of the development of weight seriation in children. Cognitive Psychology, 6, 1-40.

Black, F. 1968. A deductive question-answering system. In M. Minsky (ed.) Semantic Information Processing, MIT Press.

Block, H.D. 1970. A review of "Perceptrons: An Introduction to Computational Geometry". Information and Control, 17, 5, December, 501-522.

Brown, J.S., Rubinstein, R. and Burton, R. 1976. A reactive learning environment for computer assisted electronics instruction. BBN Report No. 3314, Bolt Beranek and Newman Inc., Cambridge, Mass.

Buchanan, B.G. and Mitchell, T.M. 1977. Model-directed learning of production rules. Stanford Heuristic Programming Project Memo HPP-77-6; Computer Science Department Report No. STAN-CS-77-597, Stanford University, March.

Burton, R. 1976. Semantic grammar: an engineering technique for constructing natural language understanding systems. BBN Report No. 3453, Bolt Beranek and Newman Inc., Cambridge, Mass.

Carbonell, J.R. 1970a. Mixed Initiative Man-Computer Instructional Dialogues. BBN Report No. 1971, Bolt Beranek and Newman Inc., Cambridge, Mass.

Carbonell, J.R. 1970b. AI in CAI: An Artificial Intelligence approach to computer assisted instruction. IEEE Transactions on Man-Machine Systems, Vol. MMS-11, No. 4.

Davis, R. In press. Knowledge acquisition in rule-based systems - knowledge about representations as a basis for system construction and maintenance. In D.A. Waterman and F. Hayes-Roth (eds.) Pattern-Directed Inference Systems, Academic Press.

Davis, R. and King, J. 1975. An Overview of Production Systems. Stanford Artificial Intelligence Laboratory Memo AIM-271; Computer Science Department Report No. STAN-CS-75-524. Stanford University, October.

di Sessa, A. 1975. ORBIT: A mini-environment for exploring orbital mechanisms. In Lecarme and Lewis (eds.) Computers in Education. Amsterdam: North Holland.

Dugdale, S. and Kibbey, D. 1976. Elementary Mathematics with Plato. Computer-based Education Laboratory, Urbana, Illinois.

Floyd, R.W. 1967. Non deterministic algorithms. Journal of the ACM, 14, 4, 636-644.

Gale, F. 1960. The theory of linear economic models. York, Pa.: Maple Press Co.

Goldberg, A. 1973. CAI: The application of theorem-proving to adaptive response analysis. Technical Report 203, Institute for Mathematical Studies in the Social Sciences, Stanford University.

Greenawalt, M. 1973. U.T. LISP manual. Computing Centre, University of Texas at Austin.

Hartley, J.R. 1973. The design and evaluation of an adaptive teaching system. International Journal of Man-Machine Studies, 5, 2.

Hartley, J.R. and Sleeman, D.H. 1971. A computer-based statistical laboratory: Some views and experiences. In D. Packham,

A. Cleary and T. Mayes (eds.) Aspects of Educational Technology, Vol. V, pp. 425-434. Pitman.

Hearn, A.C. 1970. Reduce 2 user's manual. Stanford Artificial Intelligence Project Memo AIM-133. Stanford University.

Hendrix, C., Thompson, C. and Slocum, J. 1973. Language processing via canonical verbs and semantic models. In Proceedings of Third International Joint Conference on Artificial Intelligence.

Hewitt, C. 1972. Descriptions and theoretical analysis (using schemata) of PLANNER: A language for proving theorems and manipulating models in a robot. Ph.D. Thesis, Artificial Intelligence Laboratory, MIT, Cambridge, Mass.

Howe, J.A.M. and Cassels, F. 1974. Teaching primary maths by computer. Bionics Research Report 16, School of Artificial Intelligence, University of Edinburgh.

Hughes, C.E. and Cresswell, M.S. 1972. Introduction to modal logic. London: Methuen.

Jones, T. 1971. A computer model of simple forms of learning. Ph.D. Thesis, Artificial Intelligence Laboratory, MIT, Cambridge, Mass.

Kimball, R.B. 1973. Self-optimizing computer-assisted tutoring: theory and practice. Technical Report No. 206 (Psychology and Education Series), Institute for Mathematical Studies in the Social Sciences, Stanford University, June.

Knapman, J. 1977. Artificial Learning. Ph.D. Thesis, Department of Artificial Intelligence, University of Edinburgh.

Koffman, E. 1973. Artificial Intelligence and automatic programming in CAI. Proceedings of Third International Joint Conference on Artificial Intelligence.

Lekan, H.A. 1971. Index to computer assisted instruction. New York: Harcourt Brace Inc.

Lenat, D.B. and Harris, G. 1977. Designing a rule system that searches for scientific discoveries. Technical Report, Department of Computer Science, Carnegie-Mellon University, Pittsburgh, Pa., April.

Leonard, G.B. 1968. Education and Ecstasy. New York: Delacorte Press.

McCarthy, J. and Hayes, P. 1969. Some philosophical problems from the standpoint of artificial intelligence. In B. Meltzer and D. Michie (eds.) Machine Intelligence 4, pp. 463-502. Edinburgh: Edinburgh University Press.

Minsky, M. and Papert, S. 1969. Perceptrons: An Introduction to Computational Geometry. Cambridge, Mass.: MIT Press.

Minsky, M. and Papert, S. 1972. Artificial Intelligence: Progress Report. AI Memo 252, Artificial Intelligence Laboratory, MIT, Cambridge, Mass.

Mitre Corporation. 1974. An Overview of the TICCIT Program. MITRE, McLean, Virginia.

Moses, J. 1967. Symbolic Integration. Doctoral Dissertation. MIT, Cambridge, Mass. Project MAC Report.

Newell, A. and Simon, H.A. 1972. Human Problem Solving. Prentice-Hall.

Nilsson, N.J. 1971. Problem-solving methods in artificial intelligence. McGraw-Hill.

O'Shea, T. 1973. Some experiments with an adaptive self-improving teaching system. Technical Report NL 18, Department of Computer Sciences, University of Texas at Austin.

O'Shea, T. and Sleeman, D.H. 1973. A design for an adaptive self-improving teaching system. In J. Rose (ed.) Advances in Cybernetics. Gordon & Breach.

Papert, S. 1973. Uses of technology to enhance education. LOGO Memo 8, Artificial Intelligence Laboratory, MIT, Cambridge, Mass.

Pask, G. 1972. Anti-Hodmanship: A report on the state and prospects of CAI. Richmond, Surrey: System Research Ltd., June.

Pask, G. and Scott, B.C.E. 1972. Learning strategies and individual competence. International Journal of Man-Machine Studies, 4, 217-253.

Plotkin, G. 1971. Automatic Methods of Inductive Inference. Ph.D. Thesis, School of Artificial Intelligence, University of Edinburgh.

Polya, G. 1973. How to solve it. Princeton University Press.

Post, E. 1943. Formal reductions of the general combinatorial problem. American Journal of Mathematics, 65, 147-268.

Postman, N. and Weingartner, C. 1969. Teaching as a Subversive Activity. New York: Delacorte Press.

Richards, M. 1969. BCPL reference manual. Memorandum 69/1, University Mathematical Laboratory, Cambridge, U.K.

Rosenblatt, F. 1957. The perceptron, a perceiving and recognizing automaton. Cornell Aeronautical Laboratory Report No. 85-460-1, Buffalo, N.Y.

Rychener, M.D. 1976. Production systems as a programming language for artificial intelligence applications. Department of Computer Science Report, Carnegie-Mellon University, Pittsburgh, Pa.

Samuel, A.L. 1959. Some studies in machine learning using the game of checkers. IBM Journal of Research and Development, 3, 211-229. Reprinted in E.A. Feigenbaum and J. Feldman (eds.) Computers and Thought, pp. 71-105.

Samuel, A.L. 1960. Programming computers to play games. In F.L. Alt (ed.) Advances in Computers, Vol. 1, pp. 165-192. New York: Academic Press.

Samuel, A.L. 1967. Some studies in machine learning using the game of checkers. II - Recent progress. IBM Journal of Research and Development, 11, 601-617.

Schulman, L.S. and Keisler, E.R. 1966. Learning by discovery. Chicago: Rand McNally and Co.

Self, J. 1974. Student Models in CAI. International Journal of Man-Machine Studies, 6, 2.

Shortliffe, E.H., Davis, R., Buchanan, B., Axline, B., Green, C. and Cohen, S. 1975. Computer-based consultations in clinical therapeutics: exploration and rule acquisition capabilities of the MYCIN system. Computers and Biomedical Research, 8, 303-320.

Siklòssy, L. 1970. Computer tutors that know what they teach. Proceedings Fall Joint Computer Conference, pp. 251-255.

Sleeman, D.H. 1974. A problem solving monitor for a deductive reasoning task. International Journal of Man-Machine Studies, 7, 2.

Smallwood, R.D. 1962. A Decision Structure for Teaching Machines. Cambridge, Mass.: MIT Press.

Smallwood, R.D. 1970. Optimal policy regions for computer-directed teaching systems. In W.H. Holtzman (ed.) Computer-assisted Instruction, Testing and Guidance. New York: Harper and Row.

Stansfield, J.L. 1974. Programming a dialogue teaching system. Bionics Research Report No. 25, Bionics Research Laboratory, School of Artificial Intelligence, University of Edinburgh.

Suppes, P. 1971. Computer-assisted instruction at Stanford. Technical Report No. 174, Institute for Mathematical Studies in the Social Sciences, Stanford University.

Tait, K., Hartley, J.R. and Anderson, R.C. 1973. Feedback procedures in computer-assisted arithmetic instruction. British Journal of Educational Psychology, 43, 2, 161-171.

Waterman, D.A. 1968. Machine learning of heuristics. Ph.D. Thesis, Computer Science Department, Stanford University.

Waterman, D.A. 1970. Generalization learning techniques for automating the learning of heuristics. Artificial Intelligence, 1, 121-170.

Waterman, D.A. 1977. An introduction to production systems. AISB European Newsletter, 25, 7-10, January.

Waterman, D.A. and Hayes-Roth, F. (eds.) In press. Pattern-Directed Inference Systems, Academic Press.

Waterman, D.A. and Newell, A. 1973. PAS-II. An interactive task-free version of an automatic protocol analysis system. In Proceedings of Third International Joint Conference on Artificial Intelligence.

Weissman, C. 1967. A LISP 1.5 primer. Dickenson.

Winston, P.M. 1970. Learning Structural Descriptions from Examples. Ph.D. Thesis. AI-TR-231, Artificial Intelligence Laboratory, MIT, Cambridge, Mass.

Woods, P., Hartley, J.R., Lovell, K. and Sleeman, D.H. 1969. The teaching of arithmetic using a computer-based system. In A.P. Mann and C.R. Brunstrom (eds.) Aspects of Educational Technology, Vol. III, pp. 378-385. Pitman.

Woods, P. and Hartley, J.R. 1971. Some learning models for arithmetic tasks and their use in computer-based learning. British Journal of Educational Psychology, 41, 1, 35-48, February.

Young, R.M. 1974. Production systems as models of cognitive development. Bionics Research Report No. 22, School of Artificial Intelligence, University of Edinburgh.

Young, R.M. 1976. Notes on learning. In A. Bundy et al. (eds.) To 'Artificial Intelligence', Occasional Paper No. 3, Department of Artificial Intelligence, University of Edinburgh.

Young, R.M. 1977. Mixtures of strategies in structurally adaptive production systems: examples from seriation and subtraction. D.A.I. Research Report No. 33, Department of Artificial Intelligence, University of Edinburgh.

APPENDIX I

Two protocols for the pre-test are given. Student responses are indicated by "S:"

a) Student familiar with algebraic equations:

```
HULLO I AM GOING TO TEACH YOU SOME MATH
EVERY TIME I ASK YOU A QUESTION
AFTER YOU HAVE FINISHED TYPING YOUR ANSWER
PLEASE REMEMBER TO PRESS THE RETURN KEY
ON THE RIGHT SIDE OF THE CONSOLE

WHAT IS YOUR CHRISTIAN NAME?

S:  JOHN

DO YOU UNDERSTAND EQUATIONS OF THE FORM
X - B = C

S:  YES
FINE

SOLVE FOR X  IN THE EQUATION
X - 4 = 7

S:  3
WRONG TRY AGAIN

S:  11
RIGHT

SOLVE FOR X  IN THE EQUATION
X - 5 = 10

S:  15
RIGHT

SOLVE FOR X  IN THE EQUATION
X - 7 = 14

S:  21
RIGHT

DO YOU UNDERSTAND EQUATIONS OF THE FORM
X + B = C

S:  YES
FINE

SOLVE FOR X  IN THE EQUATION
X + 3 = 6

S:  3
RIGHT

SOLVE FOR X  IN THE EQUATION
X + 5 = 10
```

```
S:  5
RIGHT

SOLVE FOR X  IN THE EQUATION
X + 8 = 15

S:  7
RIGHT

DO YOU UNDERSTAND EQUATIONS OF THE FORM
X TIMES X = B

S:  YES
FINE

SOLVE FOR X  IN THE EQUATION
X TIMES X = 64

S:  6
WRONG TRY AGAIN

S:  8
RIGHT

CONSIDER THE EQUATION
X TIMES X + 6 = X TIMES 5
*********************************

FIND VALUES FOR X  WHICH MAKE THIS EQUATION TRUE

S:  5
WRONG TRY AGAIN

S:  3
RIGHT
YES THATS RIGHT BECAUSE
3 TIMES 3 EQUALS 9
IF WE ADD 6 THAT EQUALS 15
WHICH IS EQUAL TO 3 TIMES 5
WHICH IS 15

THERE IS ANOTHER SOLUTION SEE IF YOU CAN GUESS IT

S:  9
WRONG TRY AGAIN

S:  1
NO SORRY STILL WRONG
NO THE OTHER ANSWER IS 2 BECAUSE
2 TIMES 2 EQUALS 4
IF WE ADD 6 THAT EQUALS 10
WHICH IS EQUAL TO 2 TIMES 5
WHICH IS 10
***

NOW LOOK OUT FOR RULES WHICH WILL HELP YOU SOLVE THE EQUATIONS IN
GENERAL
                         ***
```

b) Student unfamiliar with algebraic equations

```
DO YOU UNDERSTAND EQUATIONS OF THE FORM
X - B = C
```

S: YES
FINE

SOLVE FOR X IN THE EQUATION
X - 4 = 7

S: 3
WRONG TRY AGAIN

S: 2
NO SORRY STILL WRONG

LETS LOOK AT THE QUESTION ANOTHER WAY

SUPPOSE WE HAVE A BAG WITH SOME MARBLES IN
AND WE TAKE FROM THE BAG 4 MARBLES
AND THEN TIP OUT THE BAG AND FIND 7 MARBLES
HOW MANY MARBLES WERE IN THE BAG TO START WITH

S: 1
WRONG TRY AGAIN

S: 2
NO SORRY STILL WRONG

NO WHEN WE EMPTIED THE BAG WE FOUND 7 MARBLES
NOW FROM THE BAG WE HAD TAKEN 4 MARBLES
AND SO TO START WITH WE HAD 7 PLUS 4 MARBLES
MAKING THE RIGHT ANSWER 11 MARBLES

NOW SUPPOSE WE TAKE FROM THE BAG 2 MARBLES
AND THEN FIND 4
HOW MANY MARBLES WERE IN THE BAG

S: 6
RIGHT

NOW SUPPOSE WE TAKE FROM THE BAG 3 MARBLES
AND THEN FIND 7
HOW MANY MARBLES WERE IN THE BAG

S: 10
RIGHT

NOW SUPPOSE WE TAKE FROM THE BAG 6 MARBLES
AND THEN FIND 11
HOW MANY MARBLES WERE IN THE BAG

S: 17
RIGHT

NOW SUPPOSE WE TAKE FROM THE BAG 7 MARBLES
AND THEN FIND 14
HOW MANY MARBLES WERE IN THE BAG

S: 21
RIGHT

SUPPOSE WE HAVE A BAG WITH SOME MARBLES IN
AND WE ADD TO THE BAG 3 MARBLES
AND THEN TIP OUT THE BAG AND FIND 6 MARBLES
HOW MANY MARBLES WERE IN THE BAG TO START WITH

S: 3
RIGHT

NOW SUPPOSE WE ADD TO THE BAG 6 MARBLES
AND THEN FIND 11
HOW MANY MARBLES WERE IN THE BAG

S: 5
RIGHT

NOW SUPPOSE WE ADD TO THE BAG 8 MARBLES
AND THEN FIND 15
HOW MANY MARBLES WERE IN THE BAG

S: 7
RIGHT

SUPPOSE WE HAVE SOME BOYS IN A ROOM
AND EACH BOY HAS AS MANY MARBLES AS THERE ARE
BOYS IN THE ROOM AND THEY PUT ALL THEIR MARBLES TOGETHER
AND FIND THEY HAVE 64 MARBLES
HOW MANY BOYS WERE IN THE ROOM

S: 64
WRONG TRY AGAIN

S: 12
NO SORRY STILL WRONG

NO WE FOUND 64 MARBLES IN THE ROOM
SO AS THERE WERE AS MANY BOYS AS THERE WERE MARBLES
THERE MUST HAVE BEEN 8 BOYS WITH 8 MARBLES
MAKING ALTOGETHER 8 TIMES 8 MARBLES WHICH IS 64

NOW SUPPOSE WE FIND 36 MARBLES
HOW MANY BOYS WERE THERE

S: 6
RIGHT

NOW SUPPOSE WE FIND 49 MARBLES
HOW MANY BOYS WERE THERE

S: 7
RIGHT

SUPPOSE WE HAVE A GROUP OF BOYS WITH SOME MARBLES
EACH BOY HAVING AS MANY MARBLES AS THERE ARE BOYS IN THE GROUP
AND THEY ALL PUT ALL THEIR MARBLES IN A PILE
AND THEN SOMEBODY ADDS ***6*** MARBLES TO THE PILE
THEY THEN SHARE OUT THE MARBLES AND FIND THERE ARE EXACTLY ***5***
MARBLES EACH WITH NONE LEFT OVER
CAN YOU GUESS HOW MANY BOYS WERE IN THE GROUP

S: 1
WRONG TRY AGAIN

S: 2
RIGHT

YES THATS RIGHT BECAUSE

2 TIMES 2 EQUALS 4
IF WE ADD 6 THAT EQUALS 10
WHICH IS EQUAL TO 2 TIMES 5
WHICH IS 10

THERE IS ANOTHER SOLUTION SEE IF YOU CAN GUESS IT

S: 3
RIGHT
FINE CHECK THAT IS RIGHT

FINE NOW WE WILL GO ON TO THE MAIN PROBLEM

NOW LOOK OUT FOR RULES WHICH WILL HELP YOU SOLVE THE EQUATIONS IN GENERAL

APPENDIX II

Protocol of student after completing post-test

Note: The first problem is in the actual format used for all the problems. The subsequent problems have been edited in order to save space.

NOW LOOK OUT FOR RULES WHICH WILL HELP YOU SOLVE THE EQUATIONS IN GENERAL

[P1]
CONSIDER THE EQUATION X TIMES X + 4 = X TIMES 5
FIND VALUES FOR X WHICH MAKE THIS EQUATION TRUE
TRY AND GUESS THE ANSWERS

S: 4
THATS RIGHT SEE IF YOU CAN GET THE OTHER ANSWER

S: 3
NOPE

S: 2
NOPE

S: 4
YES BUT YOU HAVE GUESSED THAT CORRECTLY BEFORE

S: 5
NOPE

S: 3
NOPE

S: 6
NOPE

S: 7
NOPE

S: 8
NOPE

S: 9
NOPE

S: 10
NOPE

S: 3
NOPE

S: 0
NOPE

S: 11
NOPE

S: 12
NOPE
THAT WAS A LOT OF GUESSES
LETS TRY ANOTHER EXAMPLE

THE RIGHT ANSWERS ARE IN FACT 1 AND 4

[P2]
CONSIDER THE EQUATION X TIMES X + 7 = X TIMES 8 GUESS THE ANSWERS

S: 7
THATS RIGHT SEE IF YOU CAN GET THE OTHER ANSWER

S: 1
WELL DONE YOU HAVE GOT BOTH THE ANSWERS

[P3]
CONSIDER THE EQUATION X TIMES X + 2 = X TIMES 3 GUESS THE ANSWERS

S: 2
THATS RIGHT SEE IF YOU CAN GET THE OTHER ANSWER

S: 1
WELL DONE YOU HAVE GOT BOTH THE ANSWERS

[P4]
CONSIDER THE EQUATION X TIMES X + 5 = X TIMES 6 GUESS THE ANSWERS

S: 1
THATS RIGHT SEE IF YOU CAN GET THE OTHER ANSWER

S: 5
WELL DONE YOU HAVE GOT BOTH THE ANSWERS

[P5]
CONSIDER THE EQUATION X TIMES X + 6 = X TIMES 7 GUESS THE ANSWERS

S: 1
THATS RIGHT SEE IF YOU CAN GET THE OTHER ANSWER

S: 6
WELL DONE YOU HAVE GOT BOTH THE ANSWERS

[P6]
CONSIDER THE EQUATION X TIMES X + 8 = X TIMES 9 GUESS THE ANSWERS

S: 1
THATS RIGHT SEE IF YOU CAN GET THE OTHER ANSWER

S: 8
WELL DONE YOU HAVE GOT BOTH THE ANSWERS

[P7]
CONSIDER THE EQUATION X TIMES X + 18 = X TIMES 9 GUESS THE ANSWERS

S: WHAT HAPPENS IF I CANT

PLEASE TYPE IN A NUMBER OR TYPE STOP IF YOU WISH TO STOP

S: 1, 2
NOPE

S: 3
THATS RIGHT SEE IF YOU CAN GET THE OTHER ANSWER

S: 4, 5
NOPE

S: 6
WELL DONE YOU HAVE GOT BOTH THE ANSWERS

[P8]
CONSIDER THE EQUATION X TIMES X + 18 = X TIMES 19 GUESS THE ANSWERS

S: 3
NOPE KEEP TRYING

S: 6, 12, 2, 3
NOPE KEEP TRYING

S: 4, 5, 6, 7, 8, 9, 10
YOU ARE NOT TRYING TO THINK OF ANY RULES TO HELP YOU
WE WILL LOOK AT ANOTHER EXAMPLE
TRY TO FIND A RULE
THE RIGHT ANSWERS ARE IN FACT 1 AND 18

[P9]
CONSIDER THE EQUATION X TIMES X + 48 = X TIMES 19 GUESS THE ANSWERS

S: 1, 18, 12, 48, 19, 64, 9, 2
NOPE

S: 3
THATS RIGHT SEE IF YOU CAN GET THE OTHER ANSWER

S: 6
NOPE KEEP TRYING

S: 3
YES BUT YOU HAVE GUESSED THAT CORRECTLY BEFORE

S: 15, 18, 9, 1
NOPE

THAT WAS A LOT OF GUESSES
LETS TRY ANOTHER EXAMPLE
THE RIGHT ANSWERS ARE IN FACT 3 AND 16

[P10]
CONSIDER THE EQUATION X TIMES X + 48 = X TIMES 49 GUESS THE ANSWERS

S: 1
THATS RIGHT SEE IF YOU CAN GET THE OTHER ANSWER

S: 48
WELL DONE YOU HAVE GOT BOTH THE ANSWERS

[P11]
CONSIDER THE EQUATION X TIMES X + 9 = X TIMES 10 GUESS THE ANSWERS

S: 1
THATS RIGHT SEE IF YOU CAN GET THE OTHER ANSWER

S: 9
WELL DONE YOU HAVE GOT BOTH THE ANSWERS

[P12]
CONSIDER THE EQUATION X TIMES X + 16 = X TIMES 10 GUESS THE ANSWERS

S: 6
NOPE KEEP TRYING

S: 2
THATS RIGHT SEE IF YOU CAN GET THE OTHER ANSWER

S: 20, 32, 4, 21, 12, 13, 9, 6, 14, 13, 19, 46, 13, 1000
THAT WAS A WILD GUESS
TRY AND BE MORE SENSIBLE SUNSHINE
THE RIGHT ANSWERS ARE IN FACT 2 AND 8

[P13]
CONSIDER THE EQUATION X TIMES X + 21 = X TIMES 10 GUESS THE ANSWERS

S: 2, 8, 1
NOPE KEEP TRYING

S: 3
THATS RIGHT SEE IF YOU CAN GET THE OTHER ANSWER

S: 9, 4, 5,, 6
NOPE

S: 7
WELL DONE YOU HAVE GOT BOTH THE ANSWERS

[P14]
CONSIDER THE EQUATION X TIMES X + 24 = X TIMES 10 GUESS THE ANSWERS

S: 4
THATS RIGHT SEE IF YOU CAN GET THE OTHER ANSWER

S: 6
WELL DONE YOU HAVE GOT BOTH THE ANSWERS

[P15]
CONSIDER THE EQUATION X TIMES X + 12 = X TIMES 8 GUESS THE ANSWERS

S: 6
THATS RIGHT SEE IF YOU CAN GET THE OTHER ANSWER

S: 4, 12, 8, 9, 15, 12, 16, 13, 14, 16, 17, 18, 19, 20, 21
THAT WAS A LOT OF GUESSES
LETS TRY ANOTHER EXAMPLE
THE RIGHT ANSWERS ARE IN FACT 2 AND 6

[P16]
CONSIDER THE EQUATION X TIMES X + 15 = X TIMES 8 GUESS THE ANSWERS

S: 7, 0
NOPE KEEP TRYING

S: 3
THATS RIGHT SEE IF YOU CAN GET THE OTHER ANSWER

S: 5
WELL DONE YOU HAVE GOT BOTH THE ANSWERS

[P17]
CONSIDER THE EQUATION X TIMES X + 8 = X TIMES 6 GUESS THE ANSWERS

S: 24
NOPE

S: 2
THATS RIGHT SEE IF YOU CAN GET THE OTHER ANSWER

S: 4
WELL DONE YOU HAVE GOT BOTH THE ANSWERS

[P18]
CONSIDER THE EQUATION X TIMES X + 10 = X TIMES 7 GUESS THE ANSWERS

S: 2
THATS RIGHT SEE IF YOU CAN GET THE OTHER ANSWER

S: 5
WELL DONE YOU HAVE GOT BOTH THE ANSWERS

[P19]
CONSIDER THE EQUATION X TIMES X + 12 = X TIMES 7 GUESS THE ANSWERS

S: 3
THATS RIGHT SEE IF YOU CAN GET THE OTHER ANSWER

S: 4
WELL DONE YOU HAVE GOT BOTH THE ANSWERS

[P20]
CONSIDER THE EQUATION X TIMES X + 10 = X TIMES 11 GUESS THE ANSWERS

S: 1

THATS RIGHT SEE IF YOU CAN GET THE OTHER ANSWER

S: 10
WELL DONE YOU HAVE GOT BOTH THE ANSWERS

[P21]
CONSIDER THE EQUATION X TIMES X + 18 = X TIMES 11 GUESS THE ANSWERS

S: 2
THATS RIGHT SEE IF YOU CAN GET THE OTHER ANSWER

S: 9
WELL DONE YOU HAVE GOT BOTH THE ANSWERS

[P22]
CONSIDER THE EQUATION X TIMES X + 24 = X TIMES 11 GUESS THE ANSWERS

S: 2, 4, 1
NOPE

S: 3
THATS RIGHT SEE IF YOU CAN GET THE OTHER ANSWER

S: 6
NOPE KEEP TRYING

S: 8
WELL DONE YOU HAVE GOT BOTH THE ANSWERS

[P23]
CONSIDER THE EQUATION X TIMES X + 28 = X TIMES 11 GUESS THE ANSWERS

S: 7
THATS RIGHT SEE IF YOU CAN GET THE OTHER ANSWER

S: 4
WELL DONE YOU HAVE GOT BOTH THE ANSWERS

[P24]
CONSIDER THE EQUATION X TIMES X + 11 = X TIMES 12 GUESS THE ANSWERS

S: 1
THATS RIGHT SEE IF YOU CAN GET THE OTHER ANSWER

S: 11
WELL DONE YOU HAVE GOT BOTH THE ANSWERS

[P25]
CONSIDER THE EQUATION X TIMES X + 48 = X TIMES 16 GUESS THE ANSWERS

S: 4
THATS RIGHT SEE IF YOU CAN GET THE OTHER ANSWER

S: 12
WELL DONE YOU HAVE GOT BOTH THE ANSWERS

```
[P26]
CONSIDER THE EQUATION  X TIMES X + 48 = X TIMES 26  GUESS THE ANSWERS

S:  6
NOPE KEEP TRYING

S:  2
THATS RIGHT SEE IF YOU CAN GET THE OTHER ANSWER

S:  24
WELL DONE YOU HAVE GOT BOTH THE ANSWERS

[P27]
CONSIDER THE EQUATION  X TIMES X + 48 = X TIMES 14  GUESS THE ANSWERS

S:  3, 4
NOPE

S:  6
THATS RIGHT SEE IF YOU CAN GET THE OTHER ANSWER

S:  8
WELL DONE YOU HAVE GOT BOTH THE ANSWERS

[P28]
CONSIDER THE EQUATION  X TIMES X + 300 = X TIMES 103  GUESS THE ANSWERS

S:  3
THATS RIGHT SEE IF YOU CAN GET THE OTHER ANSWER

S:  100
WELL DONE YOU HAVE GOT BOTH THE ANSWERS

WELL DONE I THINK YOU HAVE MASTERED THIS COMPLETELY
BYE BYE CHARLIE
```

APPENDIX III

A set of production rules for the quadratic tutor

a) The elements of the state-vector

(i) CYCLE - the operation last performed by teaching program.

(ii) TIME - the number of examples administered.

(iii) NUMG - the number of guesses made by the student on the last example.

(iv) EXDETAILS - a list of the features of the last example.

(v) EXTYPE - a list of the parameters for GEN for the last example generated.

(vi) TIMES - the hypothesized current ability of the student with the times rule.

(vii) - (x) ADD, ONE, WRULE, WORULE - as in (vi) for the respective rules.

(xi) TIMESCH - the time and direction of the last change in hypothesized ability with the TIMES rule.

(xii) - (xv) ADDCH, ONECH, WRULECH, WORULECH - as in (xi) for the respective rules.

(xvi) PRETEST - the student score on the pre-test.

(xvii) REASON - the reason for termination of administration of last example.

(xviii) GOALTIME - the amount of time since the selection of the current goal.

(xix) GOAL - the current goal.

b) <u>The set of condition-action rules</u> (See note on page 167.)

```
    ( (2 B1) ( (DPRINER START1)(GOAL ≡ONE)(GEN ≡((ONE HELP)) NIL)
                (CONTINUE 1) ))
    ( (1 A3 17 I6) ( (DPRINER TERM)(STOP) ))
    ( (1 A3 2 B3) ( (DPRINER TIMETERM)(STOP) ))
    ( (1 A2 9 F1 18 J3 19 K8) ( (DPRINER SUCCTERM)(STOP) ))
    ( (1 A2 18 J3 19 K8) ( (GOAL ≡TIMES)(GEN ≡((TIMES HELP)) NIL) ))
    ( (1 A2 19 K8) ( (CONTINUE 1) ) )
    ( (1 A3 18 J4) ( (HYPALL) ))
    ( (1 A3 18 J3) ( (HYPALL) ))
    ( (1 A3 6 F1 11 G3 19 K3) ( (HYPALL) ))
    ( (1 A3 7 F1 12 G3 19 K2) ( (HYPALL) ))
    ( (1 A3 8 F1 13 G3 19 K1) ( (HYPALL) ))
    ( (1 A3 2 B2) ( (HYPALL) ))
    ( (1 A2 2 B2 10 F8) ( (DPRINER TIMETERM)(STOP)))
    ( (1 A3 19 K1) ((HYP ≡ONE) ))
    ( (1 A3 19 K2) ( (HYP ≡ADD) ))
    ( (1 A3 19 K3) ( (HYP ≡TIMES) ))
T1: ( (1 A2 6 F1 7 F1 11 G3 12 G3) ( (GOAL ≡HARD)(GEN ≡ ((WRULE HIN))
          NIL) ))
    ( (1 A2 10 F7 19 K8)( (GOAL ≡TIMES)(GEN ≡((WRULE HELP)(TIMES
          HELP)) NIL) ))
T2: ( (1 A2 6 F1 11 G3 19 K3) ( (GOAL ≡TIMESTOADD)(GEN ≡((TIMES HELP)
          (ADD HELP)) NIL) ))
T3: ( (1 A2 6 F4 7 F7 18 J3 19 K3) ( (GOAL ≡ADD)(GEN ≡((ADD HELP))
          NIL) ))
T4: ( (1 A2 6 F8 19 K3) ( (CONTINUE 1) ))
T5: ( (1 A2 7 (F8 F7) 19 K3) ( (GOAL ≡ADD)(GEN ≡((ADD HELP)) NIL) ))
T6: ( (1 A2 8 (F8 F7) 19 K3) ( (GOAL ≡ONE)(GEN ≡((ONE HELP)) NIL) ))
    ( (1 A2 7 F1 12 G3 19 K2) ( (GOAL ≡ADDTOTIMES)(GEN ≡((ADD HELP)
          (TIMES HELP)) NIL) ))
    ( (1 A2 6 F7 7 F4 18 J4 19 K2)((GOAL ≡TIMES)(GEN ≡((TIMES HELP)
          NIL) ))
    ( (1 A2 7 F8 19 K2) ( (CONTINUE 1) ))
    ( (1 A2 6 (F8 F7) 19 K2) ( (GOAL ≡TIMES)(GEN ≡((TIMES HELP))
          NIL) ))
    ( (1 A2 8 (F8 F7) 19 K2) ( (GOAL ≡ONE)(GEN ≡((ONE HELP)) NIL) ))
    ( (1 A2 8 F1 13 G3 19 K1) ( (GOAL ≡ONETOADD)(GEN ≡((ADD HELP)
          (ONE HIN)) NIL) ))
    ( (1 A2 7 F7 18 J4 19 K1) ( (GOAL ≡ADD)(GEN ≡((ADD HELP)) NIL) ))
    ( (1 A2 8 F8 19 K1) ( (CONTINUE 1) ))
    ( (1 A2 6 (F8 F7) 19 K1) ( (GOAL ≡TIMES)(GEN ≡((TIMES HELP))
          NIL) ))
    ( (1 A2 7 (F8 F7) 19 K1) ( (GOAL ≡ADD)(GEN ≡((ADD HELP)) NIL) ))
    ( (1 A2 7 F4 19 K6) ( (GOAL ≡ADD)(GEN ≡((ADD HELP)) NIL) ))
    ( (1 A2 18 J3 19 K6) ( (GOAL ≡TIMES)(GEN ≡((TIMES HELP)(ADD HIN))
          NIL) ))
    ( (1 A2 19 K6) ( (GEN ≡((TIMES HELP)(ADD HIN)) NIL)(HYP ≡TIMES)
          (GEN ≡((ADD HELP) (TIMES HIN) ) NIL)(HYP ≡ADD) ))
    ( (1 A2 6 F4 19 K7) ( (GOAL ≡TIMES)(GEN ≡((TIMES HELP)) NIL) ))
    ( (1 A2 18 J3 19 K7) ( (GOAL ≡ADD)(GEN ≡((ADD HELP)(TIMES HIN))
          NIL) ))
    ( (1 A2 19 K7)((GEN ≡((ADD HELP)(TIMES HIN)) NIL)(HYP ≡ADD)
          (GEN ≡((TIMES HELP)(ADD HIN)) NIL)(HYP ≡TIMES) ))
    ( (1 A2 8 F4 19 K4) ( (GOAL ≡ONE)(GEN ≡((ONE HELP)) NIL) ))
    ( (1 A2 18 J3 19 K4)((GOAL ≡ADD)(GEN ≡((ADD HELP)(ONE HIN)) NIL) ))
```

```
     ( (1 A2 19 K4) ( (GEN ≡((ONE HELP)(ADD HIN)) NIL)(HYP ≡ONE)
            (GEN ≡((ADD HELP)(ONE HIN)) NIL)(HYP ≡ADD) ))
T7:  ( (1 A2) ( (CONTINUE 1) ))
     ( (1 A3) ( (HYPALL) ))
     ( () ( (PRIN1 ≡FAIL) ))
```

c) <u>The set of partitions</u>

```
     (PUT ≡QUADPARTS ≡CYCLE ≡(
           (A1 EQUAL ELEMENT ≡GEN)
           (A2 EQUAL ELEMENT ≡HYP)
           (A3 EQUAL ELEMENT ≡ADMIN) ))
     (PUT ≡QUADPARTS ≡TIME ≡(
           (B1 EQUAL ELEMENT 0)
           (B2 EQUAL ELEMENT 10)
           (B3 GREATERP ELEMENT 30) ))
     (PUT ≡QUADPARTS ≡NUMG ≡(
           (C1 EQUAL ELEMENT 2)
           (C2 BETWEEN ELEMENT 11 1)
           (C3 GREATERP ELEMENT 10) ))
     (PUT ≡QUADPARTS ≡EXDETAILS ≡(
           (D1 MEMBER ≡NEW ELEMENT)
           (D2 MEMBER ≡OLD ELEMENT) ))
     (PUT ≡QUADPARTS ≡EXTYPE ≡(
        (E1 MEMBER  ≡(TIMES HELP) ELEMENT)
        (E2 MEMBER ≡(ADD HELP) ELEMENT)
        (E3 MEMBER ≡(ONE HELP) ELEMENT)
        (E4 MEMBER ≡(WRULE HIN) ELEMENT) ))
     (PUT ≡QUADPARTS ≡TIMES ≡(
           (F1 EQUAL ELEMENT ≡CERT)
           (F2 MEMBER ELEMENT ≡(CERT VPOSS POSS))
           (F3 EQUAL ELEMENT ≡DUNNO)
           (F4 MEMBER ELEMENT ≡(POSSNOT CERTNOT))
           (F5 EQUAL ELEMENT ≡CERTNOT)
           (F6 EQUAL ELEMENT ≡POSS)
           (F7 NEQ ELEMENT ≡CERT)
           (F8 MEMBER ELEMENT ≡(CERT VPOSS))
           (F9  EQUAL ELEMENT ≡NO CHECK) ))
     (PUT ≡QUADPARTS ≡TIMESCH ≡(
           (G1 EQUAL ELEMENT 0)
           (G2 EQUAL ELEMENT 1)
           (G3 GREATERP ELEMENT 2)
           (G4 EQUAL ELEMENT -1)
           (G5 LESSP ELEMENT 2) ))
     (PUT ≡QUADPARTS ≡PRETEST NIL)
     (PUT ≡QUADPARTS ≡REASON ≡(
           (I1 EQUAL ELEMENT ≡CORRECT)
           (I2 EQUAL ELEMENT ≡GUESSLIM)
           (I3 EQUAL ELEMENT ≡SYSLIM)
           (I4 EQUAL ELEMENT ≡WILDLIM)
           (I5 EQUAL ELEMENT ≡REPLIM)
           (I6 EQUAL ELEMENT ≡STOP) ))
```

```
(PUT ≡QUADPARTS ≡GOALTIME ≡(
     (J1 EQUAL ELEMENT 1)
     (J2 BETWEEN ELEMENT 4 0)
     (J3 EQUAL ELEMENT 4)
     (J3 EQUAL ELEMENT 5)
     (J4 GREATERP ELEMENT 7) ))
(PUT ≡QUADPARTS ≡GOAL ≡(
     (K1 EQUAL ELEMENT ≡ONE)
     (K2 EQUAL ELEMENT ≡ADD)
     (K3 EQUAL ELEMENT ≡TIMES)
     (K4 EQUAL ELEMENT ≡ONETOADD)
     (K5 EQUAL ELEMENT ≡ONETOTIMES)
     (K6 EQUAL ELEMENT ≡ADDTOTIMES)
     (K7 EQUAL ELEMENT ≡TIMESTOADD)
     (K8 EQUAL ELEMENT ≡HARD)
     (K9 EQUAL ELEMENT ≡EASY) ))
```

Note: The initial partitions for ADD, ONE, WRULE, WORULE are set the same as that for TIMES. Likewise the partitions for ADDCH, ONECH, WRULECH, WORULECH are the same as that for TIMESCH.

d) Example of cycle of operation for set of production rules

The state-vector: (≡HYP, 17, 8, ≡(NEW, EQ1), ≡((ONE, HELP)), CERT, CERT, VPOSS, VPOSS, CERTNOT, 4, 3, 1, 2, 7, 10, ≡CORRECT, 5, ≡ADD) when parsed with the set of partitions gives the parsed vector: (A2, NIL, C2, D1, E3, (F1, F2, F8), (F1, F2, F8), (F2, F7, F8), (F2, F7, F8), (F4, F5), G3, G3, (G2, G5), NIL, G3, NIL, I1, J3, K2). This matches the rule labelled T1 and the right-hand side is executed.

e) Parameters associated with task administration which can be used

GUESSLIM - The maximum number of guesses which the student is permitted to make.

ENCRATE - The intervals at which the student is given an encouraging remark.

REPLIM - The maximum number of repeated identical guesses before the student is moved to a new example.

SYSLIM - The maximum length of a sequence of systematic guesses before the student is given a comment on this.

WILDLIM - The highest 'wild guess' the student may make without comment.

Notes: i) The condition sides of the production rules have been written in an abbreviated form. So for example in the second rule, (1 A3 17 I6) only matches if the first element of the state-vector is an 'A3' and the seventeenth element is an 'I6'. Expressed in the notation used in chapter 3 this condition part would read

(A3 * * * * * * * * * * * * * * * I7 * *).

ii) The set of condition-action rules given here are those created after the third experiment in the series of five experiments carried out by the system.

APPENDIX IV

The set of rules for evaluating the hypothesis that the student has the TIMES rule

a) State-vector: (PAIRMULT, NUMFACT, NUMGUESSES, GAP, LASTC, HASTIMES)

b) Condition-action rules:

```
( (OR(NIL NIL 1 NIL NIL 1)
     (NIL 2 NIL NIL NIL 1)
     (3 NIL NIL 1 NIL 6)
     (NIL 3 NIL 1 NIL 6)
     (2 2 2 1 3 6)) ≡CERT)
( (OR (3 NIL 4 2 3 2)
     (NIL 3 4 2 3 2)
     (NIL NIL NIL 1 NIL 6)
     (2 2 4 2 3 2)) ≡VPOSS)
( (OR (2 NIL NIL 2 NIL NIL)
     (NIL NIL NIL NIL NIL 6)
     (NIL 2 NIL 2 NIL NIL)) ≡POSS)
( (OR (NIL NIL NIL 4 NIL NIL)
     (1 NIL NIL NIL NIL NIL)
     (NIL 1 NIL NIL NIL NIL)
     (NIL NIL 3 1 NIL NIL)) ≡CERTNOT)
( (OR (NIL NIL NIL 3 NIL NIL)
     (NIL NIL NIL NIL NIL 4)
     (NIL NIL 3 3 NIL NIL)) ≡POSSNOT)
( (NIL NIL NIL NIL NIL NIL)  DUNNO) ))
```

c) Partitions:

```
(PUT ≡HYPPARTS ≡PAIRMULT ≡(
      (1 LESSP ELEMENT 50)
      (2 GREATERP ELEMENT 50)
      (3 EQUAL ELEMENT 100) ))
(PUT ≡HYPPARTS ≡NUMFACT ≡(
      (1 LESSP ELEMENT 50)
      (2 GREATERP ELEMENT 50)
      (3 EQUAL ELEMENT 100) ))
(PUT ≡HYPPARTS ≡NUMGUESSES ≡(
      (1 EQUAL ELEMENT 2)
      (2 GREATERP ELEMENT 2)
      (3 GREATERP ELEMENT 10)
      (4 LESSP ELEMENT 6) ))
(PUT ≡HYPPARTS ≡GAP ≡(
      (1 EQUAL ELEMENT 0)
      (2 BETWEEN ELEMENT 5 -1)
      (4 GREATERP ELEMENT 1000)
      (3 GREATERP ELEMENT 4) ))
```

```
(PUT ≡HYPPARTS ≡LASTC ≡(
      (1 LESSP ELEMENT 5)
      (2 GREATERP ELEMENT 5)
      (3 GREATERP ELEMENT 10)
      (4 GREATERP ELEMENT 20)
      (5 LESSP ELEMENT 10) ))
(PUT ≡HYPPARTS ≡HASTIMES ≡(
      (1 EQUAL ELEMENT ≡CERT)
      (2 MEMBER ELEMENT ≡(CERT VPOSS POSS))
      (3 EQUAL ELEMENT ≡DUNNO)
      (4 MEMBER ELEMENT ≡(POSSNOT CERTNOT))
      (5 EQUAL ELEMENT ≡CERTNOT)
      (6 MEMBER ELEMENT ≡(VPOSS CERT)) ))
```

APPENDIX V

Task difficulty matrix for $x^2 + c = b \cdot x$ with solutions $x1$, $x2$

	NO RULE	ALL RULES	ONE	TIMES	ADD
$b < 15$	0	2	0	0	4
$1 \leq c < 8$	5	3	3	3	3
$8 \leq c < 13$	3	2	2	2	2
$13 \leq c < 31$	-3	1	1	1	0
$31 \leq c < 51$	-5	0	-1	0	0
b PRIME	0	0	0	0	2
$x1$ CONSTANT	0	2	0	3	3
$x1 = 1$	0	0	5	0	0
b CONSTANT	0	0	0	-2	5
c CONSTANT	0	3	0	4	-3
c 2 FACTORS	0	5	0	0	0
c 4 FACTORS	0	2	0	4	0
c 6 FACTORS	-1	-1	0	-1	0
c 8 FACTORS	-2	-2	0	-3	0
c 10 FACTORS	-3	-3	0	-4	0
$x1 = 100$	-1	1	0	0	1
$x1, x2 \neq 1$	0	0	-5	0	0

Note: b CONSTANT means b to be set equal to the value of b in the previous example.

APPENDIX VI

Set of assertions used by the deduction system

a) Causal assertions:

```
(POSS (DECREASE (HYPOTHESIS TESTING)) (OR(DECREASE (COMPUTER TIME))
     (INCREASE SCORES) (INCREASE (STUDENT TIME))) )
(CERT (REMOVE HYPS) (DECREASE (HYPOTHESIS TESTING)) )
(CERT (AND(EQUAL (VAR CYCLE) ADMIN)(RAISE CONTINUE))
   (DECREASE (HYPOTHESIS TESTING)) )
(CERT (AND(EQUAL SUBRULES CERT)(REPLACE (HYPALL) ((HYP ≡ONE)
   (HYP ≡ADD)(HYP ≡TIMES)))) (DECREASE (HYPOTHESIS TESTING)) )
(POSS (INCREASE (HYPOTHESIS TESTING)) (OR(INCREASE (COMPUTER TIME))
   (INCREASE SCORES) (DECREASE (STUDENT TIME)) ))
(CERT (REPLACE HYPS (HYPALL)) (INCREASE (HYPOTHESIS TESTING)) )
(POSS (REPLACE MAIN MAIN) (CHANGE SOMEGOAL) )
(POSS (REMOVE MINOR) (DECREASE (STUDENT TIME)) )
(POSS (REMOVE (GOAL ≡EASY)) (DECREASE (COMPUTER TIME)) )
(POSS (REMOVE (GOAL ≡HARD)) (DECREASE (POST SCORE)) )
(POSS (AND(BEFORE (GOAL ≡TIMES))(AFTER (GOAL≡ ONE))(INSERT
   (GOAL ≡ADD))) (CHANGE SCORES) )
(POSS (AND (REMOVE (GOAL ≡ONE))(REMOVE (GOAL ≡ONETOADD))
   (REMOVE (GOAL ≡ONETOTIMES))) (OR(DECREASE SCORES)(DECREASE TIMES)) )
(POSS (LOWER EXLIMIT) (DECREASE (STUDENT TIME)) )
(POSS (AND(AFTER (GOAL ≡EASY))(REPLACE (GOAL ≡ONE) (GOAL ≡TIMES)))
   (SHORTEN SESSIONS) )
(POSS (AND(LOWER PROGLIMIT)(PRESENT (STOP))) (SHORTEN SESSIONS) )
(CERT (SHORTEN SESSIONS) (DECREASE TIMES) )
(POSS (SHORTEN SESSIONS) (DECREASE SCORES) )
(CERT (KEEP BORED STUDENTS) (INCREASE TIMES) )
(POSS (KEEP BORED STUDENTS) (OR(DECREASE(POST SCORE))(INCREASE
   (STUDENT SCORE))) )
(CERT (AND(REPLACE (DPRINER ≡TERM)(DPRINER ≡CONTINUE))
   (REPLACE (STOP) (CONTINUE 1))) (KEEP BORED STUDENTS) )
(POSS (LOWER HYPTIME) (OR (DECREASE TIMES)(DECREASE (POST SCORE))))
(POSS (LOWER (VAR GOALTIME)) (OR (DECREASE TIMES)(DECREASE (POST
   SCORE))))
(POSS (LOWER GUESSLIM) (DECREASE (STUDENT TIME)) )
(POSS (LOWER EXLIMIT) (DECREASE SCORES) )
(POSS (AND (LOWER GUESSLIM) (INSERT (ADMIN RIGHTANS)))
   (CHANGE SCORES) )
(POSS (RAISE ENCRATE) (DECREASE (POST SCORE)) )
(POSS (DECREASE WILDLIM)(OR(DECREASE SCORES)(DECREASE (STUDENT TIME))) )
(POSS (RAISE SYSLIM) (INCREASE (STUDENT SCORE)) )
```

b) Definitional Assertions:

```
(EQUIV AFTER (NOT BEFORE))
(EQUIV ABSENT (NOT PRESENT))
(EQUIV LOWER (NOT RAISE))
(EQUIV INCREASE (NOT DECREASE))
(EQUIV CHANGE (OR INCREASE DECREASE))
(EQUIV (AND TIMES SCORES) ALLGOALS)
```

```
(EQUIV (AND (STUDENT TIME)(COMPUTER TIME)) TIMES)
(EQUIV (AND (STUDENT SCORE)(POST SCORE)) SCORES)
(EQUIV (OR (COMPUTER TIME)(STUDENT TIME)(STUDENT SCORE)
   (POST SCORE) SCORES TIMES ALLGOALS) SOMEGOAL)
(EQUIV (OR REPLIM SYSLIM GUESSLIM WILDLIM) EXLIMIT)
(EQUIV (OR (VAR TIME)(VAR PRETEST)) PROGLIMIT)
(EQUIV (OR (GOAL ≡ONETOADD)(GOAL ≡ADDTOTIMES)
   (GOAL ≡TIMESTOADD)) SWITCH)
(EQUIV (OR (GOAL ≡ADD)(GOAL ≡TIMES)(GOAL ≡ONE)) MAIN)
(EQUIV (OR (GOAL ≡HARD)(GOAL ≡EASY)) MINOR)
(EQUIV (OR MAIN MINOR SWITCH) PROGGOALS)
(EQUIV (AND(HYP ≡ONE)(HYP ≡ADD)(HYP ≡TIMES)(HYP ≡WRULE)
   (HYP ≡WORULE)(HYPALL) ) HYPS)
(EQUIV (OR (VAR ADD)(VAR ONE)(VAR TIMES)) SUBRULES)
(EQUIV (OR (VAR TIMESCH)(VAR ADDCH)(VAR ONECH)(VAR WRULECH)
   (VAR WORULECH)) HYPTIME)
```

<u>Note</u>: EQUIV is an abbreviation for DEFEQUIV (see 3.3).

APPENDIX VII

Student comments

The written post-test included a space for comments on the program. A selection of these is given below:

"I caught onto the formula. It's easy and fun."

"This is a really interesting program. I enjoyed working at the computer a lot. I learned the pattern but it took me a long time."

"I began to feel like a machine too. If I had to learn like this all the time I would get bored. I prefer personal contact. It's probably all right for science students but arts students need more stimulation."

"I enjoyed fooling with it. I wish I had one in my bedroom."

"It would be better if it could talk to you."

"It's a far-out machine."

"I think learning by this method is a good brain exercise."

"I enjoyed working on your program."

"This was a very good teaching method you have thought up."

"It was fun and I love the way you talk!"

"Why? What was the meaning of this?"

"Its really a great program, in time am sure it will help many people who are now having trouble with math! But it will take some time!"

"It makes you think of your own shortcuts to the problems."

"At first I didn't realize that x times x was really x^2, or that in fact I was working a quadradic equasion. It really is a good system because I have never really understood the quadradic equasion or how to work it."

"I think it was a fun and interesting way to learn."

"This program is very effective and I hope when I have kids they will be able to learn this way. I like it very much - it is a lot easier learning this way rather than a teacher. The learning lasts a whole lot longer. Also, computers are not prejudice!"

"The computer is real cool, I think it should be used in school."

APPENDIX VIII

The Production Rules for the pre-test administrator

a) The elements of the state-vector

(i) TYPE - the type of problem on which the student is currently being tested.

(ii) NUMWRONG - the number of examples the student has incorrectly answered.

(iii) NUMRIGHT - the number of examples the student has correctly answered.

(iv) NUMEXPND - the number of examples explained to the student.

(v) LEVEL - the level of difficulty at which the student is currently working.

(vi) LAST - the student's performance on the last example presented.

b) The actions on the right-hand sides of the rules

(i) STARTTEST - (STARTTEST NAME x) commences giving the student examples of type NAME at level of difficulty x , sets all other elements of state-vector to NIL.

(ii) STOP - stops the pre-test administrator running and passes control to the experimenter (STOP ≡SEETIM) or the main teaching program (STOP ≡SUCCESS).

(iii) TEST - administers an example of the same type and difficulty as that last administered.

(iv) HARDER, EASIER - sets the current difficulty level at a higher or lower level.

(v) EXPLAIN - explains to the student why his answer was correct (EXPLAIN ≡ANS) or incorrect (EXPLAIN ≡MISTAKE).

c) <u>The set of condition-action rules</u>

```
( (A4 * * * E2 F1)  ((STARTTEST ≡QUAD 3)) )
( (OR(A6 * * D1 *) (A6 * * * E4 *))
                    ((STOP ≡SUCCESS)) )
( (A6 * * * * *)    ((TEST)) )
( (A5 * * * * *)    ((STOP ≡SEETIM)) )
( (A1 * * * * *)    ((STARTTEST ≡SUB 3)) )
( (OR(* * * D1 * F2) (* B3 C2 * * F2) (* B4 * * F1 F2))
                    ((STOP FAIL)) )
( (A3 B1 * * E2 F1) ((STARTTEST ≡ADD 5)) )
( (A2 B1 * * E2 F1) ((STARTTEST ≡SQU 5)) )
( (A3 * * * E2 F1)  ((STARTTEST ≡ADD 3)) )
( (A2 * * * E2 F1)  ((STARTTEST ≡SQU 3)) )
( (* * * * * F1)    ((HARDER) (TEST)) )
( (* B2 C1 * * F2)  ((EXPLAIN ≡ANS) (TEST)) )
( (* * * * * F2)    ((EXPLAIN ≡MISTAKE) (EASIER)  (TEST)) )
( (* * * * * *)     ((STOP ≡SEETIM)) )
```

d) <u>The set of partitions</u>

```
(PUT ≡PREPARTS ≡TYPE ≡(
      (A1 EQUAL ELEMENT  NIL)
      (A2 EQUAL ELEMENT ≡ADD)
      (A3 EQUAL ELEMENT ≡SUB)
      (A4 EQUAL ELEMENT ≡SQU)
      (A6 EQUAL ELEMENT ≡QUAD)
      (A5 EQUAL ELEMENT ≡STOP) ))
(PUT ≡PREPARTS ≡NUMWRONG ≡(
      (B1 EQUAL ELEMENT 0)
      (B2 EQUAL ELEMENT 1)
      (B3 EQUAL ELEMENT 6)
      (B4 GREATERP ELEMENT 4) ))
(PUT ≡PREPARTS ≡NUMRIGHT ≡(
      (C1 GREATERP ELEMENT 2)
      (C2 LESSP ELEMENT 2) ))
(PUT ≡PREPARTS ≡NUMEXPND ≡(
      (D1 EQUAL ELEMENT 3)
      (D2 LESSP ELEMENT 4) ))
(PUT ≡PREPARTS ≡LEVEL ≡(
      (E1 EQUAL ELEMENT 1)
      (E4 GREATERP ELEMENT 5)
      (E2 EQUAL ELEMENT 5) ))
(PUT ≡PREPARTS ≡LAST ≡(
      (F1 EQUAL ELEMENT ≡RIGHT)
      (F2 EQUAL ELEMENT ≡WRONG) ))
```

APPENDIX IX

Glossary giving intended meanings of rare or non-standard terms

Adaptive teaching program: a program which provides individualised teaching.

Artificial student: a computer program which interacts with a teaching program in the same fashion as some class of real student.

Author language: a computer programming language designed for use in implementing teaching programs.

Causal assertion: an assertion which relates one or more actions with one or more effects possibly consequent on the execution of those actions.

Deduction procedure: an inference scheme which can be expressed as a computer program.

Definitional assertion: an assertion which names some class of the actions or effects used in the causal assertions.

Descriptive system: a general computational mechanism (such as production rules or finite state transition networks).

Discovery method: a teaching style which centres on encouraging the student to discover some concept without explicit explanation on the part of the teacher.

Educational objective: a general goal associated with some teaching situation.

Expository teaching: a teaching style which centres on presenting and explaining material to the student.

Generative teaching program: a program which uses an algorithm to generate teaching material (as opposed to those programs where each piece of teaching material is pre-stored).

Internal representation of teaching material: The descriptive system in which the teaching material of a CAI program is expressed.

Means-ends guidance rules: a mapping from possible student states of knowledge to possible teaching actions.

Mixed-initiative teaching program: a program which can reverse roles with the student and answer questions of the type normally directed to the student.

Partition: a computational rule which gives a name to some subset of the possible values of an element of a state-vector in a system of production rules.

Priming: initialising the teaching performance records of a self-improving teaching system by running it with some students.

Representation of the task: a task analysis and task difficulty model for some piece of teaching material.

Response-sensitivity: the degree to which a teaching program can adapt to the individual learning needs of its students.

Restriction: the name given by a partition (see above).

Student model: the component of a teaching program used to predict the current state of knowledge of a student.

Student-record: the database used by a teaching program in which a record of the past interactions with the student are maintained.

Task analysis: a structure imposed on some piece of teaching material (such as a partial ordering of concepts).

Task difficulty matrix: a matrix which expresses a task difficulty model.

Task difficulty model: a mapping from teaching goals to the properties of the various classes of examples which can be administered to students.

Teaching objective: an objective associated with some particular piece of teaching.

Teaching style: a mode of teaching (such as 'drill and practice' or 'expository').

Theory of instruction: a body of theory concerned with optimising the learning process.

Tutorial program: an adaptive teaching program which explains errors to the student and which it may be possible to run in mixed-initiative mode.

Tutorial strategy: those components of an adaptive teaching program which constitute the means-ends guidance rules, student model and task analysis.

Vocabulary of teaching operations: the various pieces of teaching material that can be presented to a student.

Note: Further discussion of many of the above terms can be found in Atkinson, 1972; Hartley, 1973; and Self, 1976.

SUBJECT INDEX

AUTHOR INDEX

Interdisciplinary Systems Research
Birkhäuser Verlag, Basel und Stuttgart

ISR 40
David Canfield Smith
PYGMALION: A Computer Program to Model and Stimulate Creative Thought

ISR 41
Friedrich Niehaus
Computersimulation langfristiger Umweltbelastung durch Energieerzeugung

ISR 42
Christian Konig (Herausgeber)
Energiemodelle für die Bundesrepublik Deutschland

ISR 46
Wolfgang Birkenfeld
Methoden zur Analyse von kurzen Zeitreihen

ISR 47
Takeo Kanade
Computer recognition of human faces

ISR 48
Erwin Grochla / Wolfgang Bauer / Herbert Fuchs / Helmut Lehmann / Wolfgang Vieweg
Zeitvarianz betrieblicher Systeme

ISR 49
Derek W. Bubb / Howard Thomas
Formal Methods in Policy Formulation

ISR 50
Gisela Arndt
Planung und statistische Auswertung von Computersimulationen interdependenter Modelle mit verzögerten endogenen Variablen

ISR 51
Karl A. Stroetmann (Editor)
Innovation, Economic Change and Technology Policies

ISR 52
Peter Sokolowsky
Grundlagen der Rechnertechnik mit einer Einführung in Mikroprozessoren

ISR 53
Claude Jablòn / Jean Claude Simon
Application des modèles numériques en physique

ISR 54
Reinhard Klein
Nutzenbewertung in der Raumplanung

ISR 55
Luder Bach
Methoden zur Bestimmung von Standorten und Einzugsbereichen zentraler Einrichtungen

ISR 56
Derek W. Bunn
The Synthesis of Forecasting Models in Decision Analysis

ISR 57
Constantin V. Negoita
Management Applications of System Theory

ISR 58
William S. Faught
Motivation and Intentionality in a Computer Simulation Model of Paranoia

ISR 59
Bertold Wissebach
Entscheidungen der Produktionsplanung mit einer Sammlung von BASIC-Programmen

ISR 60
Peter Diekmann / Heribert Leppert
Planspiel und Planspiel-Simulation in der Raumplanung. Text – Spiele – Literatur

ISR 61
Michael Wegener
Mensch-Maschine-Systeme für die Stadtplanung

ISR 62
Günter Halbritter
Multidimensionale Optimierung bei der Standortwahl von grosstechnischen Anlagen

ISR 63
Gerhard Knolmayer
Programmierungsmodelle für die Produktionsprogrammplanung – Ein Beitrag zur Methodologie der Modellkonstruktion

ISR 64
Michael G. Strobel
CS-PAL. Compact Statistics Programs for an Analytical Library (or rather) Computerized Statistics for the Poor and Lazy

ISR 65
Kurt Schmitz
Langfristplanung in der Energiewirtschaft – Eine Computersimulation für die BRD

ISR 66
David L. Székely
UNICODE. Ein Verfahren zur Optimierung der begrifflichen Gehirnleistung

ISR 67
Tim O'Shea
Self-Improving Teaching Systems

ISR 68
Bernhelm Booss / Mogens Niss (Editors
Mathematics and the Real World

ISR 69
Jürgen Seetzen / Rolf Krengel / Gert von Kortzfleisch (Herausgeber)
Makroökonomische Input-Output-Analysen und dynamische Modelle zur Erfassung technischer Entwicklungen

ISR 70
Edmond Bianco
Informatique Fondamentale. De la machine de turing aux ordinateurs modernes

ISR 71
Gisela Morlock-Rahn
WASSIM. Computersimulation von Wasserversorgung und Abwasserentsorgung in Verdichtungsräumen

ISR 72
Ching Y. Suen
Computational Analysis of Mandarin

MIX
Papier aus verantwortungsvollen Quellen
Paper from responsible sources
FSC® C105338

If you have any concerns about our products,
you can contact us on
ProductSafety@springernature.com

In case Publisher is established outside the EU,
the EU authorized representative is:
Springer Nature Customer Service Center GmbH
Europaplatz 3, 69115 Heidelberg, Germany

Printed by Libri Plureos GmbH
in Hamburg, Germany